BINARY STARS
A Pictorial Atlas

Dirk Terrell
Jaydeep Mukherjee
R. E. Wilson

Astronomy Department
University of Florida

With Foreword by
Slavek M. Rucinski

KRIEGER PUBLISHING COMPANY
Malabar, Florida
1992

Original Edition 1992

Printed and Published by
KRIEGER PUBLISHING COMPANY
KRIEGER DRIVE
MALABAR, FLORIDA 32950

FROM A DECLARATION OF PRINCIPLES JOINTLY ADOPTED BY A COMMITTEE OF THE AMERICAN BAR ASSOCIATION AND A COMMITTEE OF PUBLISHERS:

This publication is designed to provide accurate and authoritative information in regard to the subject matter covered. It is sold with the understanding that the publisher is not engaged in rendering legal, accounting, or other professional service. If legal advice or other expert assistance is required, the services of a competent professional person should be sought.

Library of Congress Cataloging-In-Publication Data
Terrell, Dirk.
    Binary stars : a pictorial atlas / Dirk Terrell, Jaydeep
Mukherjee, R.E. Wilson.
        p.   cm.
    Includes bibliographical references and index.
    ISBN 0-89464-041-0 (alk. paper) (cloth)
    ISBN 0-89464-698-2 (paperback)
    1. Stars, Double--Atlases. I. Mukherjee, Jaydeep. II. Wilson,
R.E. (Robert E.) 1937-   . III. Title.
QB821.T47 1992
523.8'41--dc20   91-27786
                                            CIP

10   9   8   7   6   5   4   3   2

Contents

FOREWORD

Arguably stars are the most significant component of the universe: they produce energy through thermonuclear reactions, manufacture elements leading to slow chemical evolution of the universe, and - last but not least - can possess planetary systems. Short but violent stages of stellar evolution are known to produce cosmic rays of great importance for the interstellar matter and mysterious origins of life. All clusters, galaxies, or clusters of galaxies are made of stars. However, in general, it seems to be rarely appreciated that most stars are components of binary systems: single stars like our Sun are definitely in a minority (if not entirely rare); multiple systems are also progressively less frequent for higher degrees of multiplicity. Thus the universe is mainly composed of stars in binary systems.

Binary stars are not only very common. They are also very important for astronomers since only for binaries, and for the Sun, can essential data such as masses, radii, or surface temperatures be determined. One can even say that what we know about single stars can only be inferred by similarities to already studied components of binary systems. But there is a certain price to be paid for having to study two stars in one physical system: an increased complexity. Naively one would expect that two stars in one system are about twice as complex as a single star. This may be true when components are far apart and form a well-detached system. But when separations are smaller, even approximately commensurate with sizes of stars, a host of new phenomena appear: mutual tidal interactions distort shapes of stars and induce their unusual rotation (or quench it in some cases); stars illuminate each other leading to differential heating of sides and, sometimes, to appreciable expansion and overflow. In the extreme case of contact binaries the components actually merge and form one entity with two hidden mass centers.

This book shows directly, through a multitude of graphical displays, how complex and diversified close binaries can really be. The authors present over 300 systems as they would be visible in different phases of their orbital revolutions to a moderately distant observer, perhaps as distant as we are now from our Sun. A bewildering multitude of forms and shapes must convince any reader not only of the complexity but also of the esthetic quality of this subject.

The first impression of chaos is in fact misleading. Not all combinations of shapes are permissible in the world of close binaries. One cannot put together any two stars of arbitrary sizes into one binary system. The main regulator is the Roche-lobe geometry which, although not directly visible in the figures, sets stringent limits on relative sizes of stars. A trained eye will see it, though. One can see its effects in semi-detached systems where one component has reached the limit: it is then constrained to stay at this limit by losing matter to its companion. When both components overflow their limits, as happens in contact binaries, a new requirement on the equality of potentials forces them to follow the same surface encompassing both stars.

The collection presented in this book may be best used as a convenient educational tool. This is the first attempt of which the undersigned is aware to present graphically the varied morphologies of close binary systems in such a scale and through such a multitude of cases. One must appreciate the usefulness of computer graphics for such applications, especially when coupled with powerful light-curve synthesis codes such as that of Wilson and Devinney. Also, the amount of work which went into collecting the necessary references, bringing the input data into the common system, and running all the cases must have been colossal. All this is hidden in the simplicity and the visual appeal of the presentation. What remains for the reader is to enjoy the world of close binaries, a fascinating world which although well understood has never been seen by human eyes.

Institute for Space and Terrestrial Science
and York University
Toronto, Ontario

SLAVEK M. RUCINSKI

About the Pictorial Atlas

In some well-publicized areas of astronomy and astrophysics, it is not unusual to meet persons from everyday life who are reasonably familiar with the essentials, due largely to programs on educational television which provide good information on cosmology, solar system exploration, black holes, quasars, and so on. However the field of interacting binary stars is the arena of professional astronomers, with some involvement by seriously dedicated amateur observers. Persons who have an interest in astronomy (casual, active, or intense), but are not active in interacting binary star research, frequently lack perspective and knowledge of the fundamentals. The situation is better with widely separated, non-interacting binaries. There the interested amateur may have a reasonable background as a result of observing visual binaries, which are among the most rewarding targets for a backyard telescope. Yet familiarity with binary stars by the typical educated and astronomically curious person need not be limited to wide binaries. Many of the principles which guide interacting binary research are understandable from a common sense viewpoint. Certain of these principles can be illustrated nicely in pictures - especially through a substantial collection of pictures, to be perused at leisure. Although we cannot go to these remote star systems and take photographs, we can produce computer simulations based on published estimates of their dimensions and other properties, and that is what has been done in the **Pictorial Atlas**. A glossary has been included in order to supplement the pictures with straightforward explanations of key terminology. The **Atlas** is intended for binary star professionals also, and the next section is an introduction for research and teaching applications.

Those who like astronomy, and believe they might like interacting binary stars if they got to know them, should begin from the following basis. Most of those points of light in the sky are binary star systems, or even multiple systems (3,4,5,6. . . stars). Single stars are the exception, not the rule. Binary systems are bound by gravitation into stable, enduring configurations. The range of separations is enormous - from actual contact up to far larger than our planetary system. The **Pictorial Atlas** shows only examples with small separations - roughly up to that between the solar system's innermost planet, Mercury, and the Sun. These are the ones which tend to interact most strongly, and in the greatest variety of ways. To be seen in the pictures are the effects of tides, rotation, magnetic starspots, exchange of matter, overcontact, eclipse circumstances, and eccentric orbits. Still other effects, such as non-uniform surface brightness ("gravity darkening"), aspect-dependence of surface brightness ("limb darkening"), and heating of each star by the other ("reflection effect") can be seen in the illustrated light curves. However this is not a textbook in binary star astronomy, for explanations of how and why all of these effects operate would exceed a reasonable space for a book of this kind. Essential background on binary star morphology, evolution, matter exchange, and related topics may be obtained from R.E.W.'s article "Binary Stars, a Look at Some Interesting Developments" in Mercury magazine (Sep./Oct., 1974), which has been reprinted in "The New Astronomy and Space Science Reader" (ed. J.C. Brandt and S.P. Maran, W.H. Freeman and Co., 1977).

The Pictorial Atlas for Research and Teaching

A primary purpose of the **Pictorial Atlas** is to provide visualizations of astrophysically interesting binary stars. Phenomena which might remain hidden in a stream of numbers can become quite obvious in a picture. For example, one can see the dramatic effect of hot spots on far-ultraviolet light curves.

The computer-generated pictures and light curves were made with a version of the Wilson-Devinney program (ApJ **166**, 605; ApJ **234**, 1054) which was modified to interface with a plotting package. The surfaces of the WD model stars are surfaces of constant potential energy (equipotentials), which are computed under the assumption of the extended Roche model. The model applies to semi-detached binaries such as the Algols, overcontact systems such as the W UMa stars, and detached systems. The model can be applied to systems with eccentric orbits and with non-synchronous rotation, and a few such binaries are included in the **Atlas**. We have selected the systems to illustrate a wide variety of morphological and astrophysical types.

The **Atlas** could be used in the classroom to illustrate close binary morphology. For example, not only the overcontact condition, but degrees of overcontact are seen as one flips through the pages, and the marginal contact of the W-type W UMa systems can be contrasted with the strong overcontact of the A-types. Variations among Algol-type systems show that some Algols are quite different, in absolute and relative dimensions, from a familiar example such as Algol itself. The effect of fast rotation can be striking in Algol-type binaries with accreting primaries, and especially in systems which are in or near to the double contact condition. The **Atlas** should be effective in providing intuition for solution of light curves.

The pictures show each system at phases 0.00, 0.05, 0.10, 0.15, 0.25, 0.35, 0.40, 0.45, and 0.50, starting at the upper left and moving left to right and top to bottom. The stars move with respect to their common center of mass which has a fixed location in the nine "frames". The systems have been arranged into groups based on the size of the semi-major axis of the relative orbit (i.e. $a = a_1 + a_2$). The circle to the right of the pictures illustrates the size of the Sun to the same scale as the binary, so as to indicate the absolute size scale. Systems within a given group are scaled by a fixed factor so that intercomparison of size within the group also is easy. The groups are defined as follows:

I	$1.50 \, R_\odot < a \le 2.25 \, R_\odot$
II	$2.25 \, R_\odot < a \le 3.38 \, R_\odot$
III	$3.38 \, R_\odot < a \le 5.06 \, R_\odot$
IV	$5.06 \, R_\odot < a \le 7.60 \, R_\odot$
V	$7.60 \, R_\odot < a \le 11.39 \, R_\odot$
VI	$11.39 \, R_\odot < a \le 17.08 \, R_\odot$
VII	$17.08 \, R_\odot < a \le 25.63 \, R_\odot$
VIII	$25.63 \, R_\odot < a \le 38.44 \, R_\odot$
IX	$38.44 \, R_\odot < a \le 57.37 \, R_\odot$

Below the pictures are the computed light curves in the infrared (16000Å), visible (5500Å), and ultraviolet (2000Å) regions of the spectrum, from top to bottom. The vertical lines just above the light curves indicate phase 0.00 (left) and 0.50 (right). All of the light curves have been normalized to unity at phase 0.25. The horizontal lines to the left and right of the light curves indicate the fifty percent light level (relative to that at phase 0.25) for the infrared, visible, and ultraviolet light curves from top to bottom. Values for the limb darkening coefficients were estimated from the tables of Carbon and Gingerich (in <u>Theory and Observation of Normal Stellar Atmospheres</u>, ed. O. Gingerich, Cambridge MA: M.I.T., p. 377). Bolometric albedos were set equal to 1.0 and 0.5 for radiative and convective envelopes respectively, except in cases where the original authors found significantly different values. Gravity darkening exponents were 1.0 for radiative envelopes and 0.3 for convective envelopes. A few systems have small amounts of estimated third light in the visible bandpass, but we did not include it in computing the light curves because the values are unknown for the ultraviolet and infrared light curves, and we wished to maintain consistency among the light curves. All of the light curves were computed assuming Planckian radiation, so the ultraviolet (and perhaps the infrared) light curves should not be taken too literally. However, those curves may point to some phenomena of general interest.

A number of physical conditions, such as lobe-filling, provide natural restrictions on binary star configurations. These conditions are implemented in the WD program by allowing the specification of certain **modes** of operation. Under each mode the program applies particular constraints among parameter values, except in mode 0, which has no constraints. This results in values which are necessarily consistent with the assumed physical conditions. Modes 1 and 3 are intended for overcontact binaries, so the surface potentials of the two stars are required to be the same. Mode 3 permits a temperature discontinuity on the connecting neck, while mode 1 does not. Mode 2 is for detached binaries, with no constraints on potentials, but luminosities are required to be consistent with surface temperatures (not the case in mode 0). Modes 4 and 5 are for the semi-detached condition, with the potential of star 1 set to the lobe-filling value for mode 4 and that of star 2 set for lobe-filling in mode 5. Mode 6 is for double-contact binaries, with the potentials of both stars set for lobe-filling. Mode -1 is for eclipsing X-ray binaries, where the duration of eclipse is known from the X-ray observations, but there are no detectable optical eclipses. Solution of optical light curves as those of an ellipsoidal variable is constrained to be consistent with the known eclipse duration. The mode used for the light curves of each binary in the **Pictorial Atlas** is given in the list of references.

In the upper right corner of each page is the group number and name of the binary. It is imperative to keep certain qualifiers in mind. Parameter values have been taken from the astronomical literature available to us. The systems of the **Atlas** have published photometric solutions which employ a variety of analysis methods. Because different models have different parameters, it was often necessary to translate parameters into those of the WD model. One common example is the conversion of relative radii of solutions from D.B. Wood's WINK program, which is based on ellipsoidal stars, into the potentials of WD. This was done by iterating the surface potentials until the side radius was equal to the "b" radius of the WINK solution. For most WINK solutions of semi-detached systems, we used the published mass ratio and allowed the lobe-filling star to assume the dimensions of the lobe. However, in some cases where this procedure led to significant inconsistencies in star size, we used the given dimensions and iterated to find the necessary mass ratio. Another warning arises because several systems have only one published solution and that solution is for a morphological type for which that model was not designed. The most common example of this is the application of an ellipsoid model to a semi-detached system.

The following information is listed at the bottom of each page:

a	semi-major axis of the relative orbit ($a_1 + a_2$) in solar radii
e	eccentricity of the orbit
ω	longitude of periastron in degrees
P	period in days
i	inclination of the orbit in degrees
T_1, T_2	mean surface temperatures of the components in Kelvins
r(pole)	relative star radius perpendicular to the plane of the orbit, with the orbital semi-major axis as the unit
r(point)	relative star radius along the line of centers directed toward the other star (undefined for contact binaries and listed as -1.000 in this case)
r(side)	relative star radius in the plane of the orbit perpendicular to the line of centers
r(back)	relative star radius along the line of centers directed away from the other star
Ω_1, Ω_2	modified surface potential for each star
q	mass ratio of secondary to primary. The primary star is, by definition, the one for which superior conjunction occurs at or near phase zero.
V_γ	systemic velocity in km sec^{-1}
F_1, F_2	ratio of (surface) spin angular speed to mean orbital angular speed for each component

Not all of the systems have published radial velocity data, and thus a direct determination of the absolute dimensions was not always possible. For semi-detached binaries with at least one component apparently on the main sequence we estimated the mass of that star, based on its spectral type, and used the photometric mass ratio to estimate the total system mass. The semi-major axis of the relative orbit was then computed via Kepler's third law. Systems for which this was done have the value of *a* quoted to one decimal place and, of course, have no listed value for the systemic velocity. One should take the absolute dimensions of the systems which lack radial velocity work for what they are: nothing more than reasonable guesses. A small number of systems have published velocities used by the authors of the photometric solution, but we were unable to examine the original velocity data because the publications are not in our holdings. Typically, the authors of the photometric solutions used the velocities to compute absolute dimensions, but did not list a value of V_γ. In these cases we have listed the value of *a* to two or more decimal places as given by the authors of the photometric solution, but we list no value of V_γ. For serious work, one should consult the original papers. The list of references contains the source of the photometric solution and the source of the velocity data, if any. For references on radial velocity data and judgments of reliability, we have relied heavily on the Eighth Catalogue of the Orbital Elements of Spectroscopic Binary Systems compiled by A.H. Batten, J.M. Fletcher, and D.G. MacCarthy at the Dominion Astrophysical Observatory. However, we also consulted the original papers for most binaries. Systems for which this catalog was the only radial velocity reference are denoted by "(DAO)" placed after the original velocity reference. In these cases we were not able to consult the original paper and we took the elements from the DAO catalog. For photometric solutions, the catalog by B. Cester, B. Fedel, G. Giuricin, F. Mardirossian, F. Mezzetti, and F. Predolin was often helpful (MSAI **50**; 1979).

In conclusion we emphasize that the purpose of the **Pictorial Atlas** is not to provide a compilation of elements that might be used for statistical analysis. The **Atlas** should never be so used because it is, in all likelihood, rife with selection effects. Furthermore, for some binaries the absolute dimensions are based only on reasonable estimates, rather than on radial velocity measures. We have reservations about some of the published solutions and the actual parameters of some systems may be substantially different from those listed. We can only provide assurance that those parameters, when put into the WD program, yield the results shown. Perhaps the most satisfying use of the **Atlas** will be its help in uncovering implausible configurations, leading to new observations and improved understanding of binaries.

We have tried to eliminate transcription errors, but some are bound to occur. Some errors of judgment are also inevitable, and we do not claim that the solutions shown are the best available in all cases, although we have made a reasonable effort to select good solutions. We would very much appreciate having errors of any kind brought to our attention.

Group I
Systems with
$1.50\,R_\odot < a \leqslant 2.25\,R_\odot$

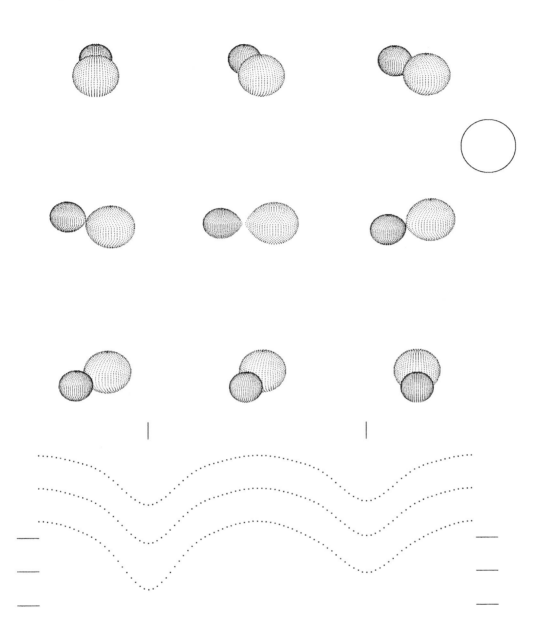

a= 1.94 $R_\odot$ r_1(pole)=0.297 r_2(pole)=0.412

e= 0.000 r_1(point)=−1.000 r_2(point)=−1.000

ω= − − − r_1(side)=0.310 r_2(side)=0.437

P=0^d.2678 r_1(back)=0.341 r_2(back)=0.464

i=$70°$.9 Ω_1=5.272 V_γ= − − − −

T_1=5800 K Ω_2=5.272 F_1=1.00

T_2=5553 K   q=1.997 F_2=1.00

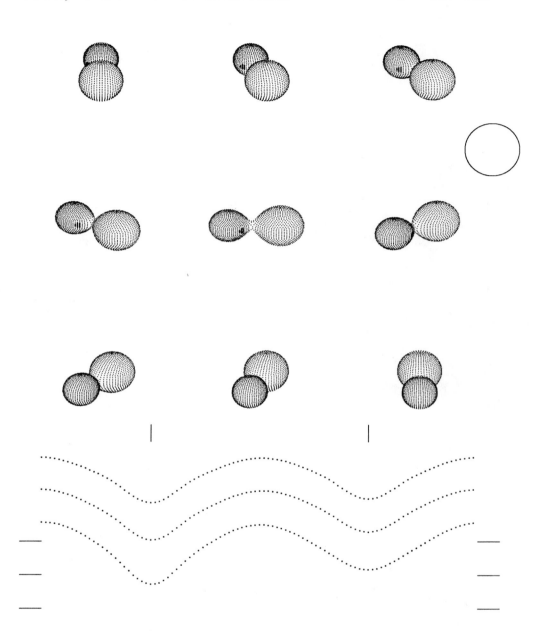

a= 1.9 R$_\odot$ r$_1$(pole)=0.404   r$_2$(pole)=0.326

e= 0.000   r$_1$(point)=−1.000   r$_2$(point)=−1.000

ω= − − −   r$_1$(side)=0.428   r$_2$(side)=0.342

P=0^d.2827   r$_1$(back)=0.461   r$_2$(back)=0.379

i=64°.9 Ω_1=3.058   V$_\gamma$= − − −

T$_1$=4825 K Ω_2=3.058   F$_1$=1.0

T$_2$=5164 K   q=0.625   F$_2$=1.0

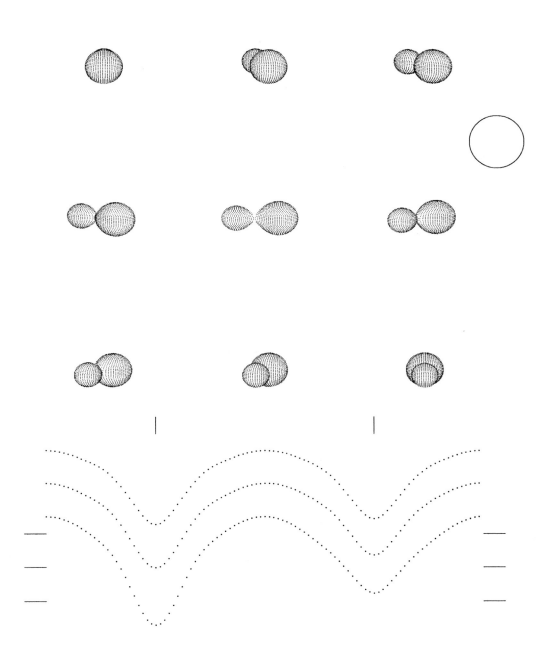

a=1.52 R$_\odot$ r$_1$(pole)=0.308   r$_2$(pole)=0.422

e= 0.000   r$_1$(point)=−1.000   r$_2$(point)=−1.000

ω= − − −   r$_1$(side)=0.323   r$_2$(side)=0.450

P=0^d.2337   r$_1$(back)=0.360   r$_2$(back)=0.481

i=82°.0 Ω_1=5.160   V$_\gamma$=−3.0 km sec^{-1}

T$_1$=4400 K Ω_2=5.160   F$_1$=1.00

T$_2$=4215 K   q=2.000   F$_2$=1.00

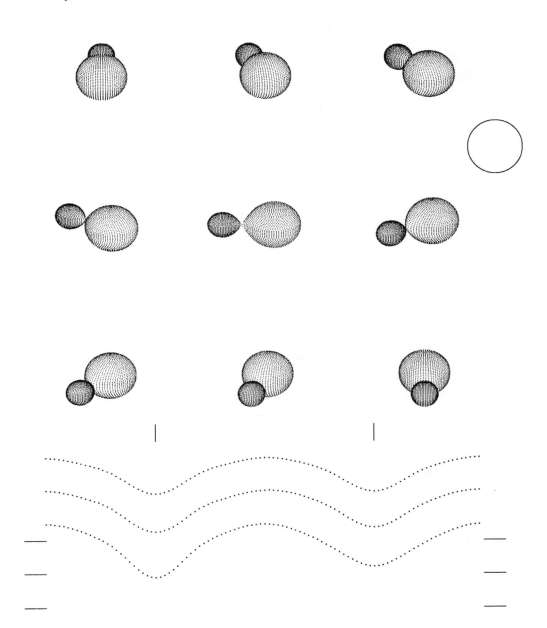

a= 1.87 R$_\odot$	r$_1$(pole)=0.254	r$_2$(pole)=0.463
e= 0.000	r$_1$(point)=0.370	r$_2$(point)=0.630
ω= –––	r$_1$(side)=0.265	r$_2$(side)=0.499
P=0^d.2783	r$_1$(back)=0.297	r$_2$(back)=0.524
i=65°.0	Ω_1=7.490	V$_\gamma$=−8.0 km sec^{-1}
T$_1$=5200 K	Ω_2=7.490	F$_1$=1.0
T$_2$=5000 K	q=3.67	F$_2$=1.0

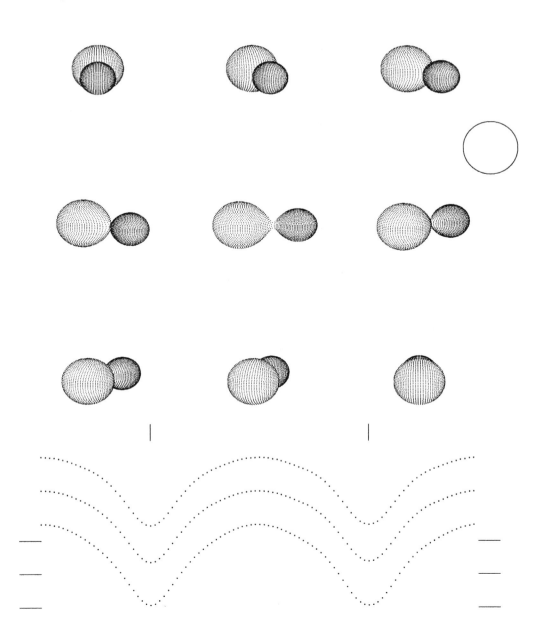

a= 2.1 $R_\odot$	r_1(pole)=0.422	r_2(pole)=0.298
e= 0.000	r_1(point)=−1.000	r_2(point)=−1.000
ω= −−−	r_1(side)=0.449	r_2(side)=0.311
P=0^d.2857	r_1(back)=0.478	r_2(back)=0.345
i=$80°$.0	Ω_1=2.802	V_γ= −−−
T_1=5250 K	Ω_2=2.802	F_1=1.0
T_2=5250 K	q=0.47	F_2=1.0

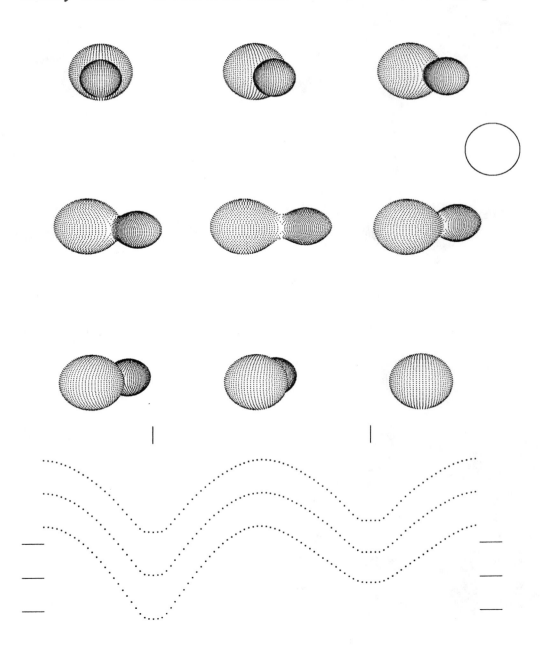

$$a=2.2\ R_\odot \qquad r_1(\text{pole})=0.480 \qquad r_2(\text{pole})=0.322$$

$$e=\ 0.000 \qquad r_1(\text{point})=-1.000 \qquad r_2(\text{point})=-1.000$$

$$\omega=\ --- \qquad r_1(\text{side})=0.527 \qquad r_2(\text{side})=0.345$$

$$P=0^{\text{d}}.3189 \qquad r_1(\text{back})=0.574 \qquad r_2(\text{back})=0.451$$

$$i=83°.9 \qquad \Omega_1=2.415 \qquad V_\gamma=\ ---$$

$$T_1=5800\ K \qquad \Omega_2=2.415 \qquad F_1=1.00$$

$$T_2=6113\ K \qquad q=0.370 \qquad F_2=1.00$$

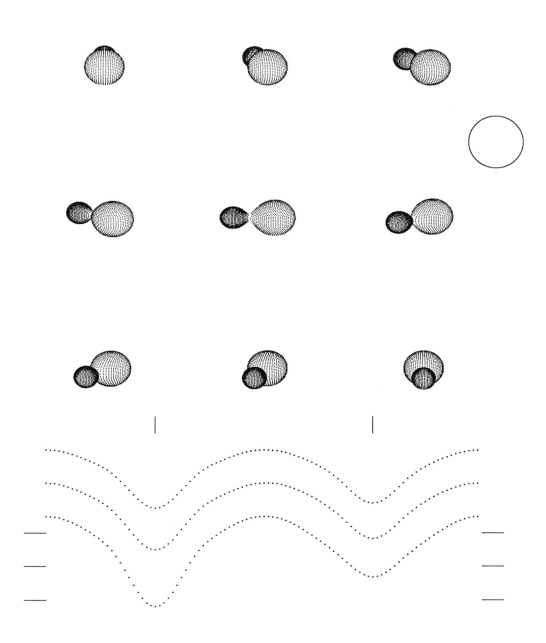

a= 1.50 R$_\odot$ r$_1$(pole)=0.279   r$_2$(pole)=0.453

e= 0.000   r$_1$(point)=−1.000   r$_2$(point)=−1.000

ω= ———   r$_1$(side)=0.292   r$_2$(side)=0.487

P=0^d.237   r$_1$(back)=0.331   r$_2$(back)=0.516

i=75°.2 Ω_1=6.395   V$_\gamma$=−53.0 km sec^{-1}

T$_1$=5400 K Ω_2=6.395   F$_1$=1.00

T$_2$=5078 K   q=2.920   F$_2$=1.00

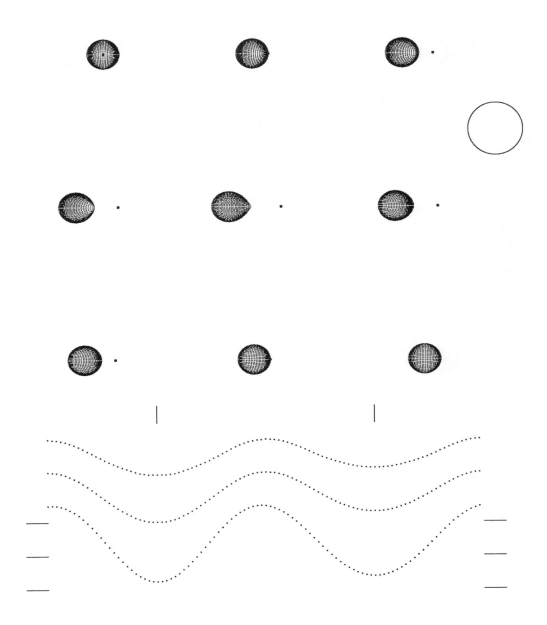

a= 1.9 R$_\odot$ r$_1$(pole)=0.300   r$_2$(pole)= ---

e= 0.000   r$_1$(point)=0.429   r$_2$(point)= ---

ω= ---   r$_1$(side)=0.313   r$_2$(side)= ---

P=0^d.2182   r$_1$(back)=0.345   r$_2$(back)= ---

i=90°.0 Ω_1=5.252   V$_\gamma$= ---

T$_1$=6700 K Ω_2= ---   F$_1$=1.00

T$_2$= ---   q=2.0   F$_2$=1.00

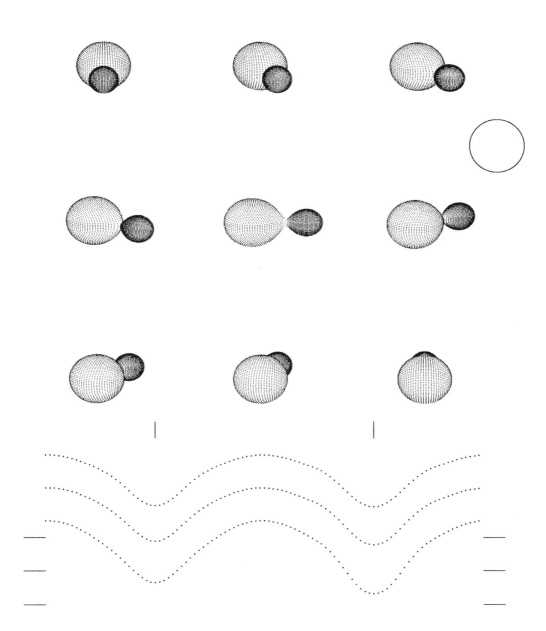

a= 1.96 $R_\odot$	r_1(pole)=0.466	r_2(pole)=0.262
e= 0.000	r_1(point)=−1.000	r_2(point)=−1.000
ω= − − −	r_1(side)=0.504	r_2(side)=0.273
P=0^d.2923	r_1(back)=0.530	r_2(back)=0.310
i=$74°$.42	Ω_1=2.397	V_γ=−60.1 km sec^{-1}
T_1=5980 K	Ω_2=2.397	F_1=1.0
T_2=6164 K	q=0.280	F_2=1.0

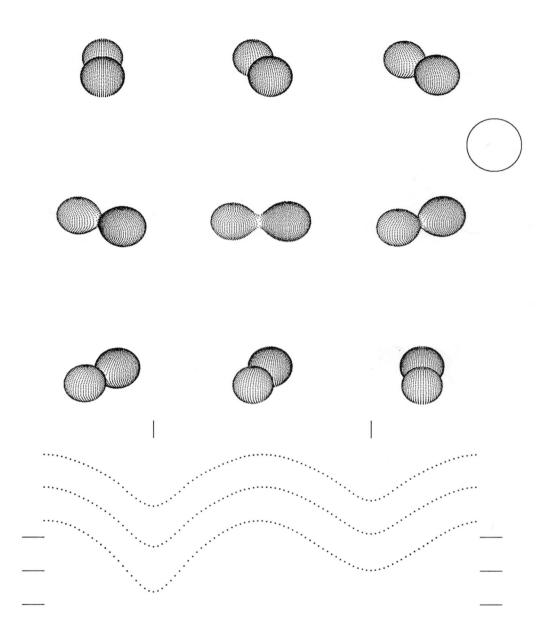

a= 1.98 $R_\odot$ r_1(pole)=0.355 r_2(pole)=0.384

e= 0.000 r_1(point)=−1.000 r_2(point)=−1.000

ω= −−− r_1(side)=0.375 r_2(side)=0.407

P=0^d.2841 r_1(back)=0.415 r_2(back)=0.445

i=69°.00 Ω_1=3.935 $V_\mp$51.8 km sec^{-1}

T_1=4600 K Ω_2=3.935 F_1=1.00

T_2=4230 K   q=1.190 F_2=1.00

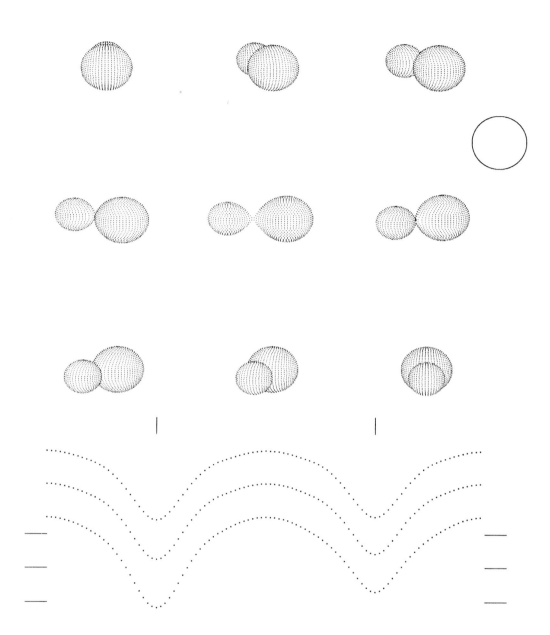

a=2.12 R$_\odot$	r$_1$(pole)=0.300	r$_2$(pole)=0.414
e= 0.000	r$_1$(point)=−1.000	r$_2$(point)=−1.000
ω= ---	r$_1$(side)=0.314	r$_2$(side)=0.440
P=0^d.4234	r$_1$(back)=0.346	r$_2$(back)=0.468
i=80°.89	Ω_1=5.230	V$_\gamma$=35.0 km sec^{-1}
T$_1$=5800 K	Ω_2=5.230	F$_1$=1.00
T$_2$=5650 K	q=1.990	F$_2$=1.00

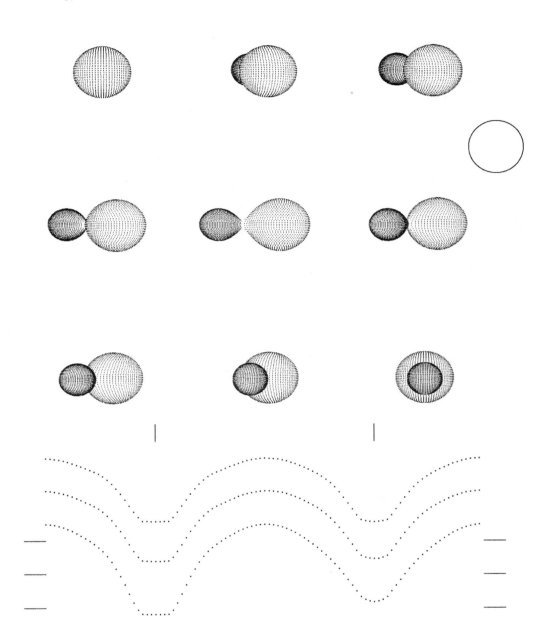

$a=$ 2.2 $R_\odot$ $r_1(\text{pole})=0.278$ $r_2(\text{pole})=0.446$

$e=$ 0.000 $r_1(\text{point})=-1.000$ $r_2(\text{point})=-1.000$

$\omega=$ ——— $r_1(\text{side})=0.290$ $r_2(\text{side})=0.479$

$P=0^d.2804$ $r_1(\text{back})=0.327$ $r_2(\text{back})=0.506$

$i=88^\circ.4$ $\Omega_1=6.284$ $V_\gamma=$ ———

$T_1=5721$ K $\Omega_2=6.284$ $F_1=1.0$

$T_2=5500$ K $q=2.80$ $F_2=1.0$

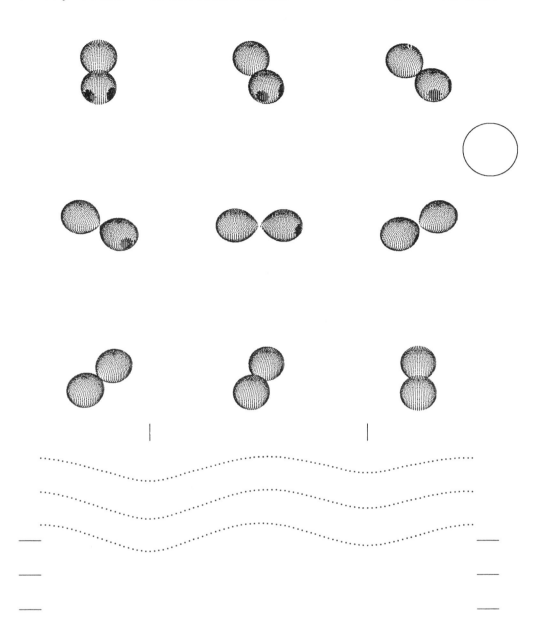

a= 1.76 R$_\odot$ r$_1$(pole)=0.360   r$_2$(pole)=0.349

e= 0.000   r$_1$(point)=0.470   r$_2$(point)=0.491

ω= ———   r$_1$(side)=0.378   r$_2$(side)=0.366

P=0$^{\rm d}$.2612   r$_1$(back)=0.407   r$_2$(back)=0.398

i=49°.9 Ω_1=3.643   V$_\gamma$=−4.3 km sec^{-1}

T$_1$=4500 K Ω_2=3.619   F$_1$=1.0

T$_2$=4305 K   q=0.92   F$_2$=1.0

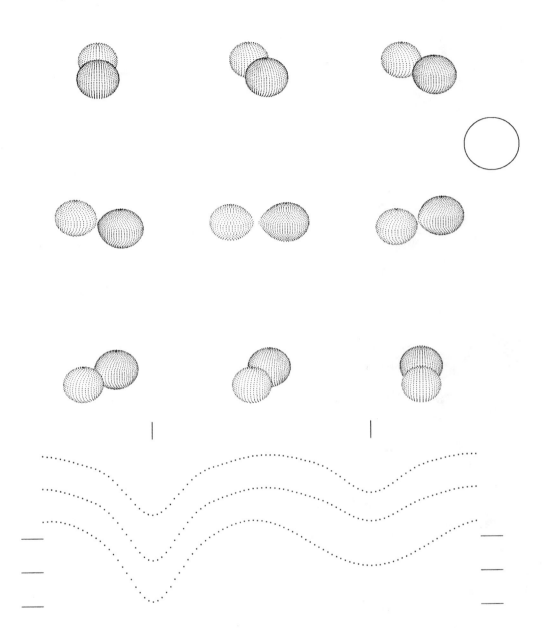

a=2.03 $R_\odot$ r_1(pole)=0.336 r_2(pole)=0.364

e= 0.000 r_1(point)=0.415 r_2(point)=0.510

ω= ——— r_1(side)=0.351 r_2(side)=0.383

P=0^d.5116 r_1(back)=0.377 r_2(back)=0.414

i=70°.05 Ω_1=4.015 V_γ=30.6 km sec^{-1}

T_1=7250 K Ω_2=3.911 F_1=1.00

T_2=4757 K   q=1.100 F_2=1.00

Group II
Systems with
$2.25\ R_\odot < a \leqslant 3.38\ R_\odot$

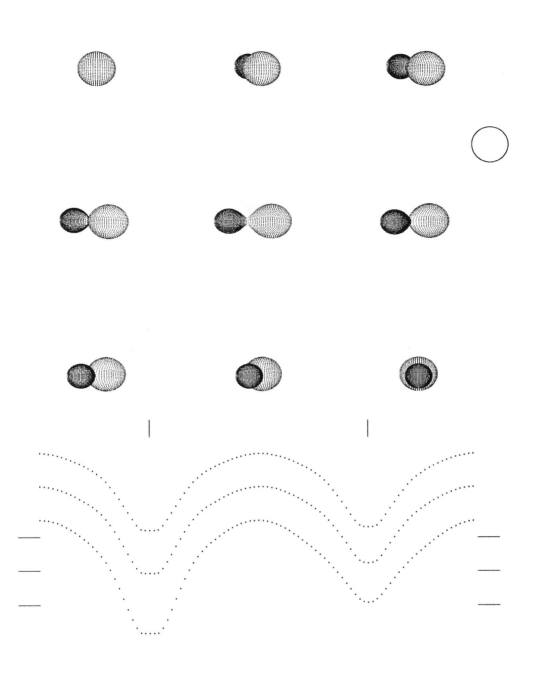

$$a=\ 2.308\ R_{\odot}\qquad r_1(\text{pole})=0.306\qquad r_2(\text{pole})=0.423$$

$a=\ 2.308\ R_{\odot}$	$r_1(\text{pole})=0.306$	$r_2(\text{pole})=0.423$
$e=\ 0.000$	$r_1(\text{point})=-1.000$	$r_2(\text{point})=-1.000$
$\omega=\ ---$	$r_1(\text{side})=0.320$	$r_2(\text{side})=0.451$
$P=0^{d}.3319$	$r_1(\text{back})=0.358$	$r_2(\text{back})=0.482$
$i=86^{\circ}.80$	$\Omega_1=5.216$	$V_{\gamma}=-24.6\ \text{km sec}^{-1}$
$T_1=5821\ K$	$\Omega_2=5.216$	$F_1=1.0$
$T_2=5450\ K$	$q=2.037$	$F_2=1.0$

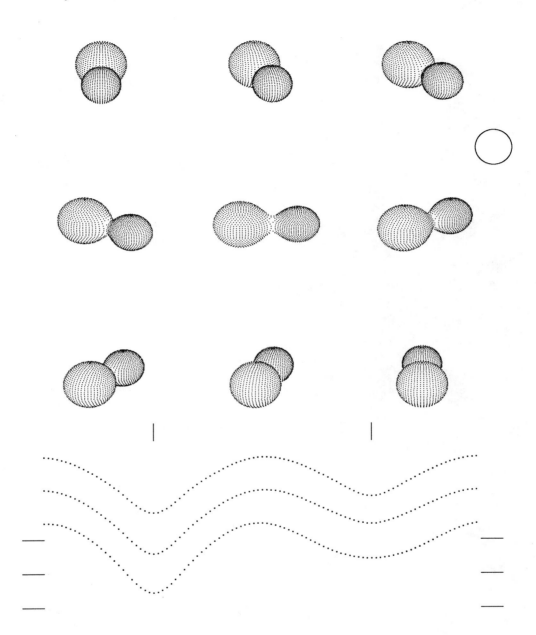

a=3.1 R$_\odot$ r$_1$(pole)=0.424   r$_2$(pole)=0.334

e= 0.000   r$_1$(point)=−1.000   r$_2$(point)=−1.000

ω= −−−   r$_1$(side)=0.454   r$_2$(side)=0.353

P=0^d.4628   r$_1$(back)=0.494   r$_2$(back)=0.403

i=68°.49 Ω_1=2.892   V$_\gamma$= −−−

T$_1$=6200 K Ω_2=2.892   F$_1$=1.00

T$_2$=4552 K   q=0.579   F$_2$=1.00

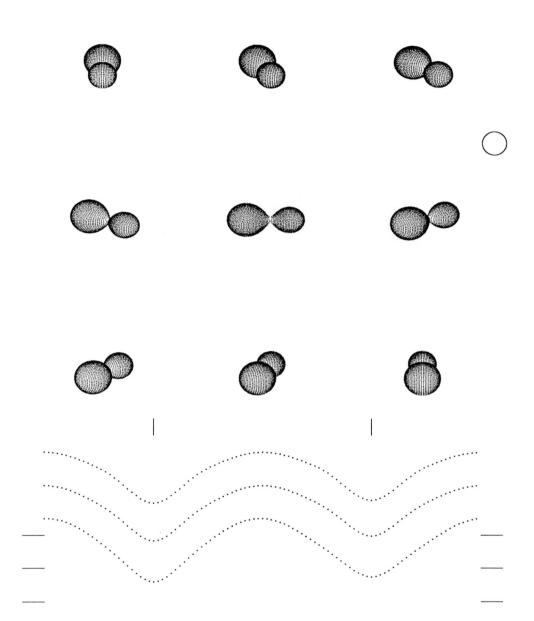

a=3.5 R$_\odot$ r$_1$(pole)=0.405   r$_2$(pole)=0.318

e= 0.000   r$_1$(point)=−1.000   r$_2$(point)=−1.000

ω= – – –   r$_1$(side)=0.429   r$_2$(side)=0.332

P=0^d.6483   r$_1$(back)=0.461   r$_2$(back)=0.368

i=69°.49 Ω_1=3.014   V$_\gamma$=+15.0 km sec^{-1}

T$_1$=7800 K Ω_2=3.014   F$_1$=1.00

T$_2$=7340 K   q=0.59   F$_2$=1.00

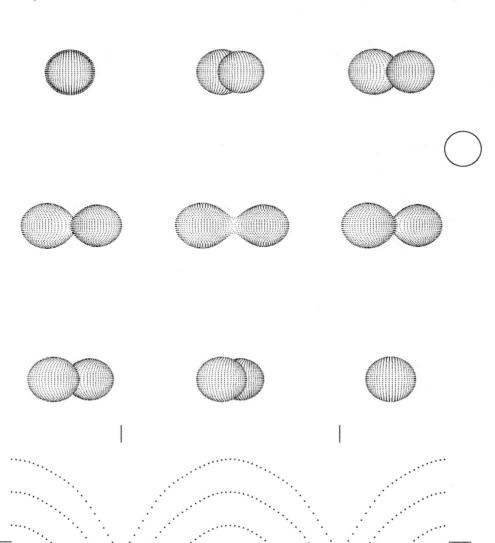

a= 3.325 R$_\odot$ r$_1$(pole)=0.388   r$_2$(pole)=0.359

e= 0.000   r$_1$(point)=−1.000   r$_2$(point)=−1.000

ω= ---   r$_1$(side)=0.412   r$_2$(side)=0.380

P=0^d.5068   r$_1$(back)=0.451   r$_2$(back)=0.422

i=90°.0 Ω_1=3.364   V$_\gamma$=−45.2 km sec^{-1}

T$_1$=5700 K Ω_2=3.364   F$_1$=1.0

T$_2$=5635 K   q=0.843   F$_2$=1.0

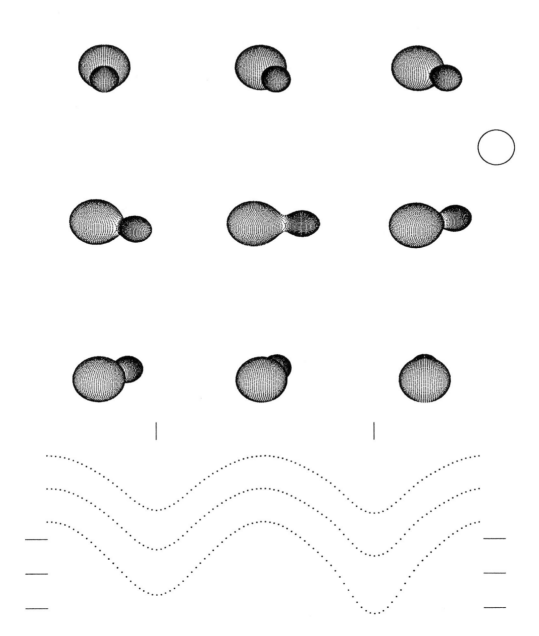

$a=2.6\ R_\odot$ $r_1(\text{pole})=0.494$ $r_2(\text{pole})=0.288$

$e=\ 0.000$ $r_1(\text{point})=-1.000$ $r_2(\text{point})=-1.000$

$\omega=\ ---$ $r_1(\text{side})=0.543$ $r_2(\text{side})=0.305$

$P=0^{\text{d}}.4060$ $r_1(\text{back})=0.580$ $r_2(\text{back})=0.388$

$i=74^\circ.9$ $\Omega_1=2.265$ $V_\gamma=\ ---$

$T_1=5600\ K$ $\Omega_2=2.265$ $F_1=1.00$

$T_2=5919\ K$ $q=0.270$ $F_2=1.00$

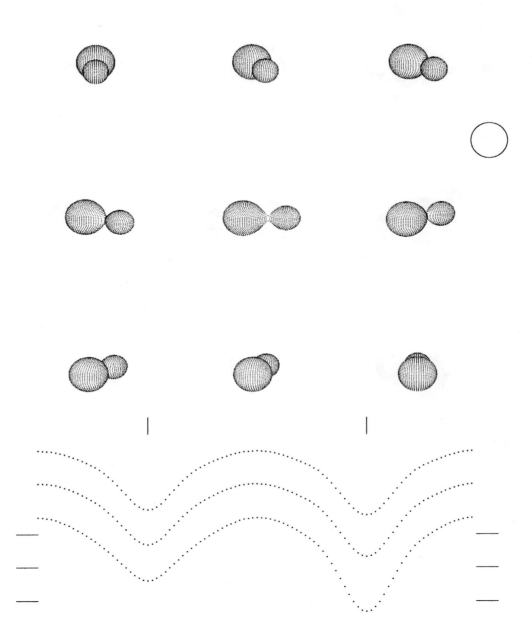

$a=2.32\ R_\odot$ $r_1(\text{pole})=0.428$ $r_2(\text{pole})=0.296$

$e=\ 0.000$ $r_1(\text{point})=-1.000$ $r_2(\text{point})=-1.000$

$\omega=\ ---$ $r_1(\text{side})=0.457$ $r_2(\text{side})=0.309$

$P=0^d.3172$ $r_1(\text{back})=0.486$ $r_2(\text{back})=0.345$

$i=77°.1$ $\Omega_1=2.746$ $V_\gamma=-38.7\ \text{km sec}^{-1}$

$T_1=5300\ K$ $\Omega_2=2.746$ $F_1=1.00$

$T_2=5623\ K$ $q=0.447$ $F_2=1.00$

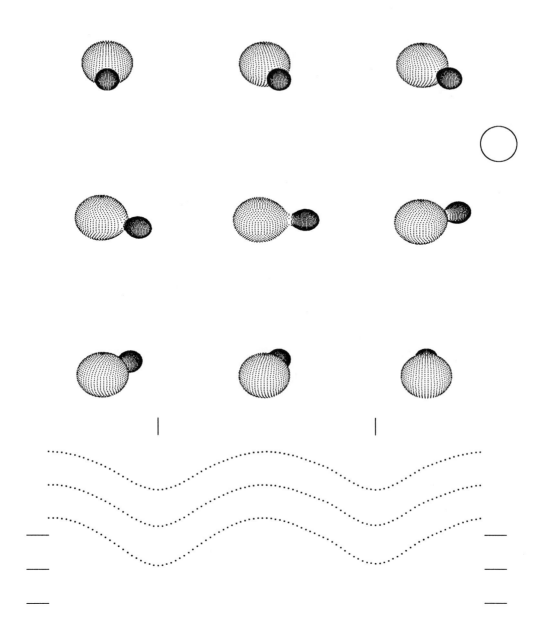

a=2.5 $R_\odot$ r_1(pole)=0.508 r_2(pole)=0.244

e= 0.000 r_1(point)=−1.000 r_2(point)=−1.000

ω= --- r_1(side)=0.559 r_2(side)=0.257

P=0^d.3705 r_1(back)=0.586 r_2(back)=0.309

i=$68°$.38 Ω_1=2.134 V_γ= ---

T_1=7200 K Ω_2=2.134 F_1=1.00

T_2=7102 K   q=0.185 F_2=1.00

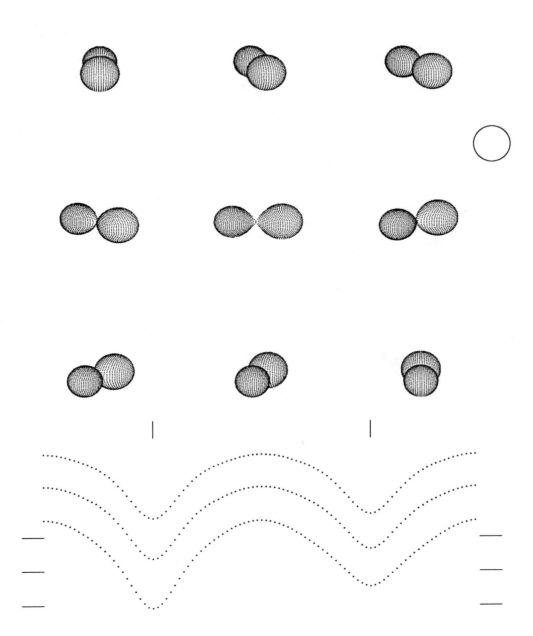

$a = 2.7\ R_\odot$ $r_1(pole) = 0.336$ $r_2(pole) = 0.379$

$e = 0.000$ $r_1(point) = -1.000$ $r_2(point) = -1.000$

$\omega = $ ——— $r_1(side) = 0.352$ $r_2(side) = 0.399$

$P = 0^d.3299$ $r_1(back) = 0.384$ $r_2(back) = 0.430$

$i = 75°.9$ $\Omega_1 = 4.214$ $V_\gamma = $ ———

$T_1 = 5826\ K$ $\Omega_2 = 4.214$ $F_1 = 1.00$

$T_2 = 5520\ K$ $q = 1.300$ $F_2 = 1.00$

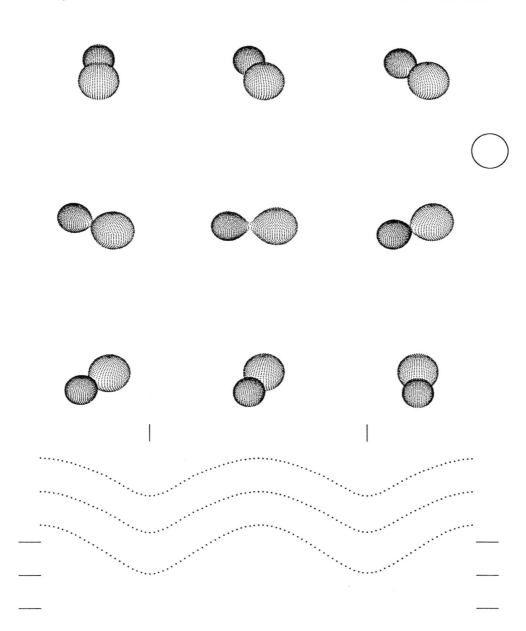

a= 2.61 $R_\odot$ r_1(pole)=0.318 r_2(pole)=0.404

e= 0.000 r_1(point)=−1.000 r_2(point)=−1.000

ω= −−− r_1(side)=0.333 r_2(side)=0.428

P=0^d.383 r_1(back)=0.368 r_2(back)=0.459

i=62°.8 Ω_1=4.741 V_γ=26.6 km sec^{-1}

T_1=6400 K Ω_2=4.741 F_1=1.00

T_2=6338 K   q=1.678 F_2=1.00

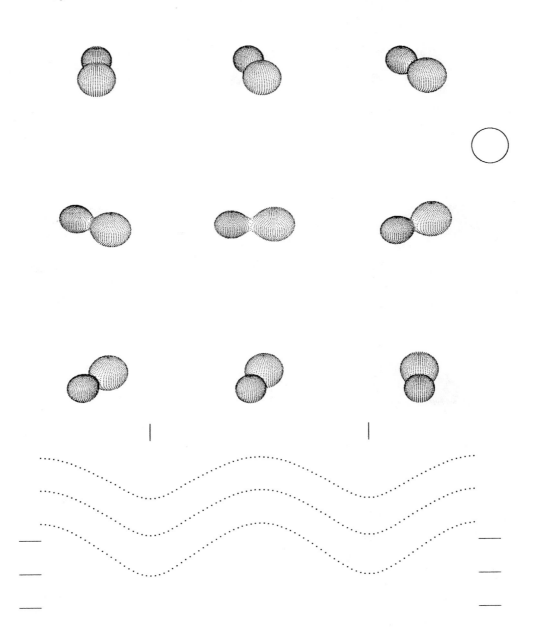

a=2.32 $R_{\odot}$ r_1(pole)=0.340 r_2(pole)=0.421

e= 0.000 r_1(point)=−1.000 r_2(point)=−1.000

ω= — — — r_1(side)=0.359 r_2(side)=0.451

P=0^d.3604 r_1(back)=0.410 r_2(back)=0.492

i=$62°$.94 Ω_1=4.494 V_{γ}=10.0 km sec^{-1}

T_1=6500 K Ω_2=4.494 F_1=1.00

T_2=6416 K   q=1.640 F_2=1.00

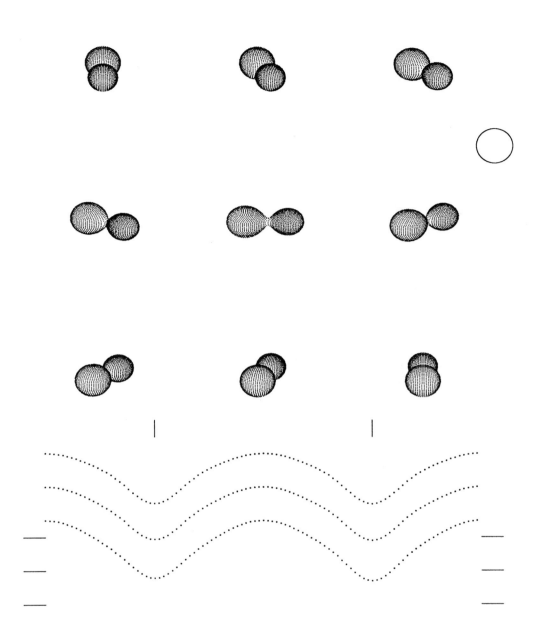

a=2.32 $R_\odot$ r_1(pole)=0.394 r_2(pole)=0.335

e= 0.000 r_1(point)=−1.000 r_2(point)=−1.000

ω= −−− r_1(side)=0.418 r_2(side)=0.352

P=0^d.5808 r_1(back)=0.451 r_2(back)=0.389

i=69°.1 Ω_1=3.189 V_γ=50.0 km sec^{-1}

T_1=5800 K Ω_2=3.189 F_1=1.00

T_2=5846 K   q=0.701 F_2=1.00

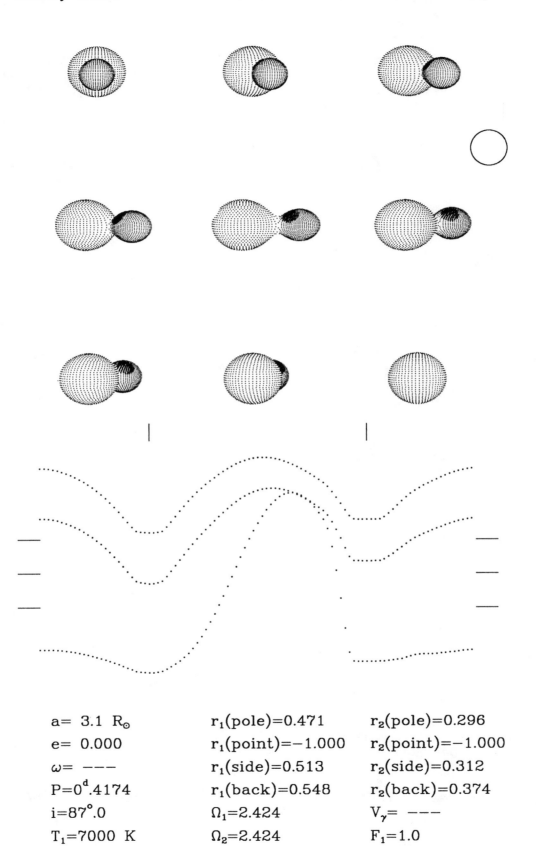

a= 3.1 $R_\odot$	r_1(pole)=0.471	r_2(pole)=0.296
e= 0.000	r_1(point)=−1.000	r_2(point)=−1.000
ω= –––	r_1(side)=0.513	r_2(side)=0.312
P=0^d.4174	r_1(back)=0.548	r_2(back)=0.374
i=$87°$.0	Ω_1=2.424	V_γ= –––
T_1=7000 K	Ω_2=2.424	F_1=1.0
T_2=6000 K	q=0.335	F_2=1.0

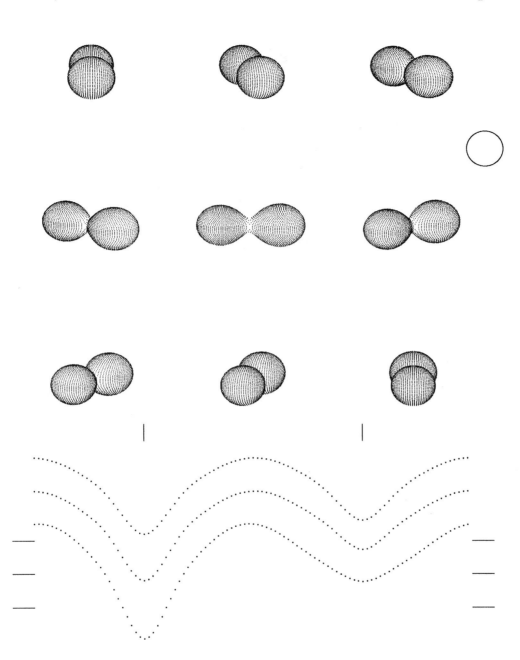

a= 3.2 R$_\odot$	r$_1$(pole)=0.360	r$_2$(pole)=0.383
e= 0.000	r$_1$(point)=−1.000	r$_2$(point)=−1.000
ω= −−−	r$_1$(side)=0.380	r$_2$(side)=0.406
P=0^d.4244	r$_1$(back)=0.420	r$_2$(back)=0.444
i=76°.9	Ω_1=3.862	V$_\gamma$= −−−
T$_1$=5900 K	Ω_2=3.862	F$_1$=1.0
T$_2$=5030 K	q=1.15	F$_2$=1.0

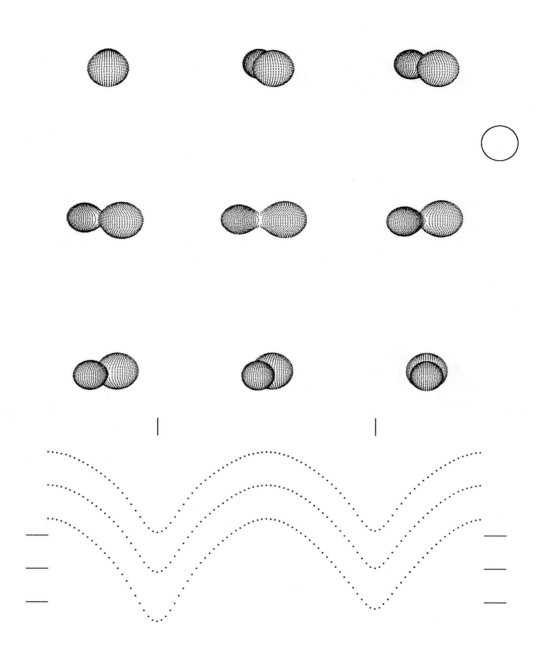

$a=2.45\ R_\odot$ $r_1(\text{pole})=0.344$ $r_2(\text{pole})=0.428$

$e=\ 0.000$ $r_1(\text{point})=-1.000$ $r_2(\text{point})=-1.000$

$\omega=\ ---$ $r_1(\text{side})=0.366$ $r_2(\text{side})=0.460$

$P=0^d.3169$ $r_1(\text{back})=0.424$ $r_2(\text{back})=0.505$

$i=83°.72$ $\Omega_1=4.743$ $V_\gamma=20.0\ \text{km sec}^{-1}$

$T_1=5600\ K$ $\Omega_2=4.743$ $F_1=1.00$

$T_2=5470\ K$ $q=1.664$ $F_2=1.00$

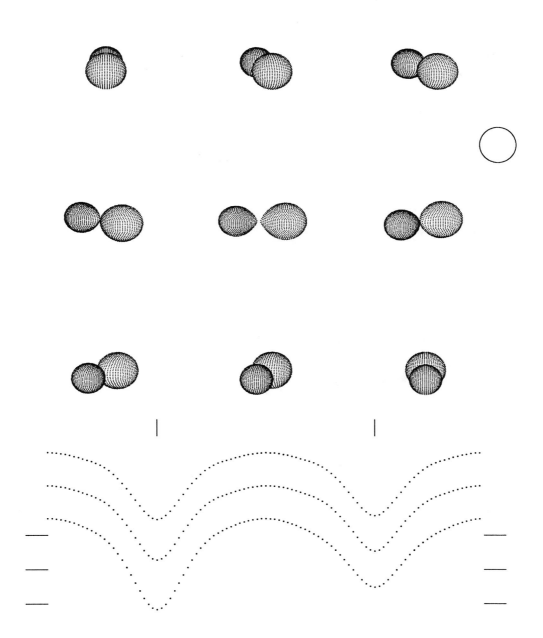

$$a=2.7\ R_\odot \qquad r_1(\text{pole})=0.320 \qquad r_2(\text{pole})=0.388$$

$$e=\ 0.000 \qquad r_1(\text{point})=0.425 \qquad r_2(\text{point})=0.508$$

$$\omega=\ \text{---} \qquad r_1(\text{side})=0.334 \qquad r_2(\text{side})=0.409$$

$$P=0^d.3408 \qquad r_1(\text{back})=0.365 \qquad r_2(\text{back})=0.438$$

$$i=78^\circ.45 \qquad \Omega_1=4.556 \qquad V_\gamma=\ \text{---}$$

$$T_1=5610\ K \qquad \Omega_2=4.556 \qquad F_1=1.00$$

$$T_2=5393\ K \qquad q=1.504 \qquad F_2=1.00$$

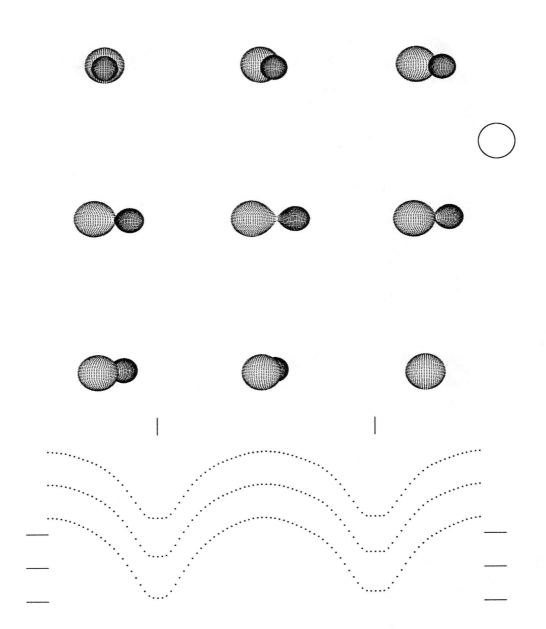

a= 2.37 R$_\odot$ r$_1$(pole)=0.428   r$_2$(pole)=0.292

e= 0.000   r$_1$(point)=−1.000   r$_2$(point)=−1.000

ω= ---   r$_1$(side)=0.457   r$_2$(side)=0.305

P=0^d.339   r$_1$(back)=0.485   r$_2$(back)=0.340

i=86°.0 Ω_1=2.735   V$_\gamma$=−1.8 km sec^{-1}

T$_1$=5564 K Ω_2=2.735   F$_1$=1.00

T$_2$=5500 K   q=0.436   F$_2$=1.00

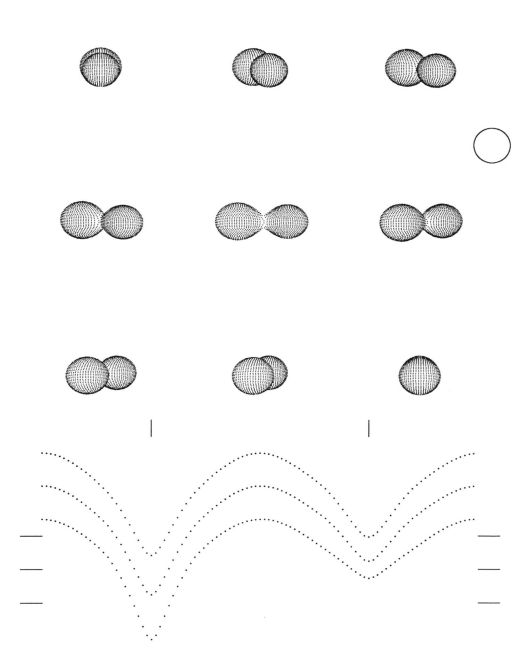

$$a=2.7\ R_\odot \qquad r_1(\text{pole})=0.397 \qquad r_2(\text{pole})=0.361$$

$$e=\ 0.000 \qquad r_1(\text{point})=-1.000 \qquad r_2(\text{point})=-1.000$$

$$\omega=\ ---\qquad r_1(\text{side})=0.422 \qquad r_2(\text{side})=0.383$$

$$P=0^d.3881 \qquad r_1(\text{back})=0.465 \qquad r_2(\text{back})=0.429$$

$$i=86°.04 \qquad \Omega_1=3.272 \qquad V_\gamma=\ ---$$

$$T_1=5600\ K \qquad \Omega_2=3.272 \qquad F_1=1.00$$

$$T_2=4937\ K \qquad q=0.807 \qquad F_2=1.00$$

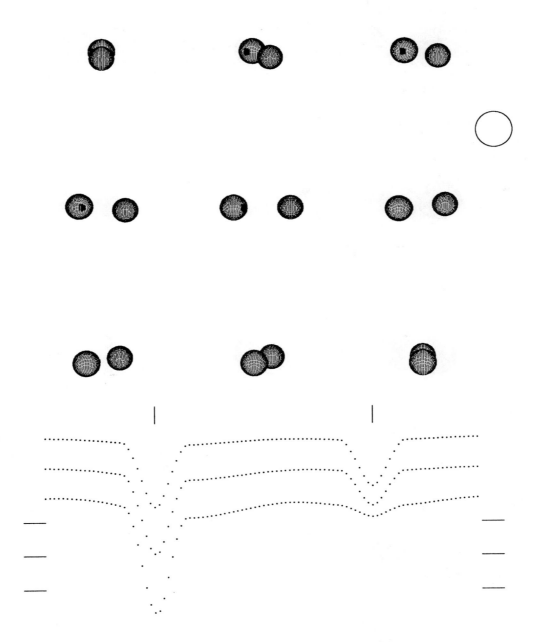

a= 3.13 R$_\odot$ r$_1$(pole)=0.235   r$_2$(pole)=0.221

e= 0.000   r$_1$(point)=0.245   r$_2$(point)=0.228

ω= ———   r$_1$(side)=0.238   r$_2$(side)=0.223

P=0^d.6311   r$_1$(back)=0.243   r$_2$(back)=0.226

i=83°.0 Ω_1=5.227   V$_\gamma$=+6 km sec^{-1}

T$_1$=5200 K Ω_2=5.509   F$_1$=1.0

T$_2$=4400 K   q=1.0   F$_2$=1.0

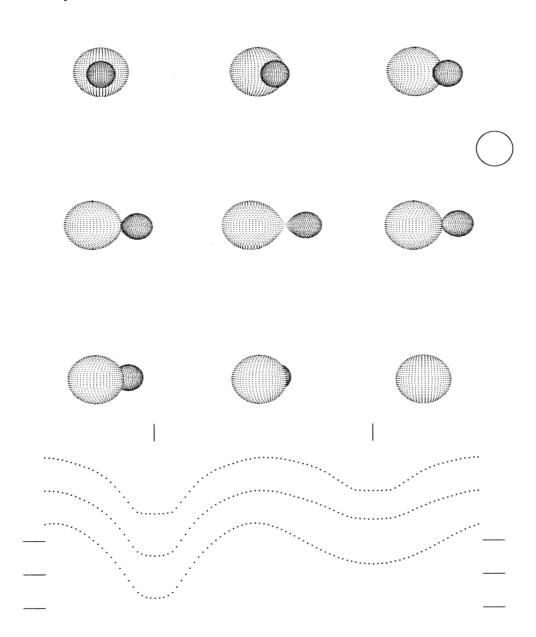

$$a=\ 3.0\ R_{\odot} \qquad r_1(\text{pole})=0.467 \qquad r_2(\text{pole})=0.251$$

$$e=\ 0.000 \qquad r_1(\text{point})=0.634 \qquad r_2(\text{point})=0.368$$

$$\omega=\ \text{---} \qquad r_1(\text{side})=0.504 \qquad r_2(\text{side})=0.261$$

$$P=0^d.3330 \qquad r_1(\text{back})=0.528 \qquad r_2(\text{back})=0.294$$

$$i=87°.4 \qquad \Omega_1=2.377 \qquad V_\gamma=\ \text{---}$$

$$T_1=8770\ K \qquad \Omega_2=2.376 \qquad F_1=1.00$$

$$T_2=5017\ K \qquad q=0.260 \qquad F_2=1.00$$

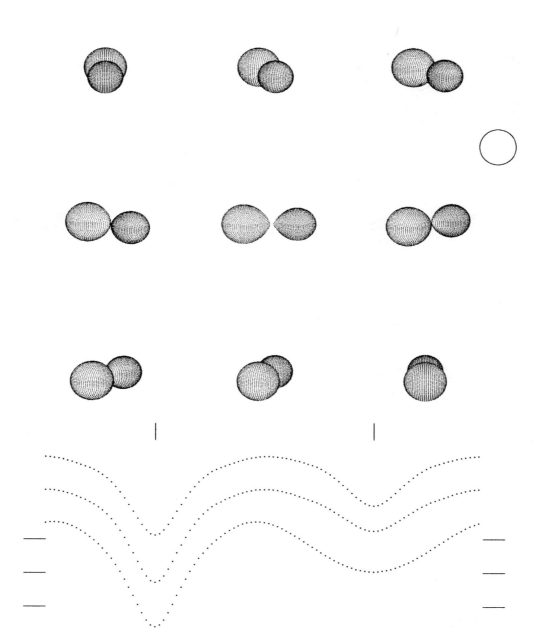

a= 2.78 $R_{\odot}$ r_1(pole)=0.384 r_2(pole)=0.322

e= 0.000 r_1(point)=0.485 r_2(point)=0.433

ω= — — — r_1(side)=0.405 r_2(side)=0.337

P=0^d.5509 r_1(back)=0.432 r_2(back)=0.369

i=$78°$.67 Ω_1=3.220 V_γ= −16 km sec^{-1}

T_1=9070 K Ω_2=3.179 F_1=1.00

T_2=5597 K   q=0.664 F_2=1.00

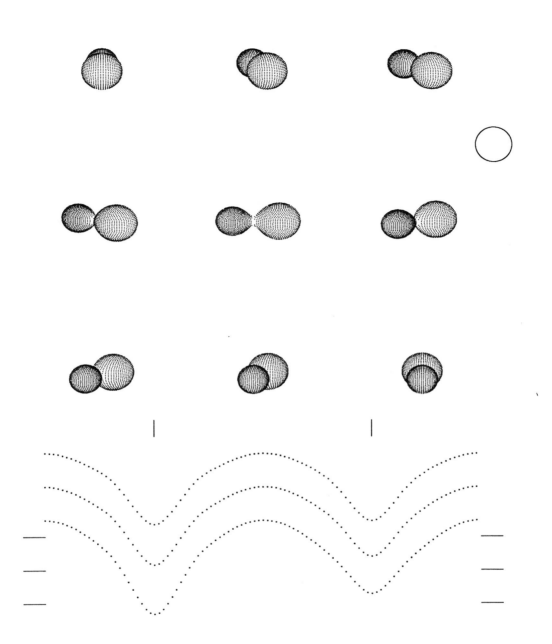

a= 2.54 $R_\odot$	$r_1(pole)=0.322$	$r_2(pole)=0.411$
e= 0.000	$r_1(point)=-1.000$	$r_2(point)=-1.000$
ω= — — —	$r_1(side)=0.338$	$r_2(side)=0.438$
P=$0^d.322$	$r_1(back)=0.376$	$r_2(back)=0.470$
i=$79^\circ.4$	$\Omega_1=4.738$	V_γ= — — —
T_1=5600 K	$\Omega_2=4.738$	$F_1=1.00$
T_2=5397 K	q=1.715	$F_2=1.00$

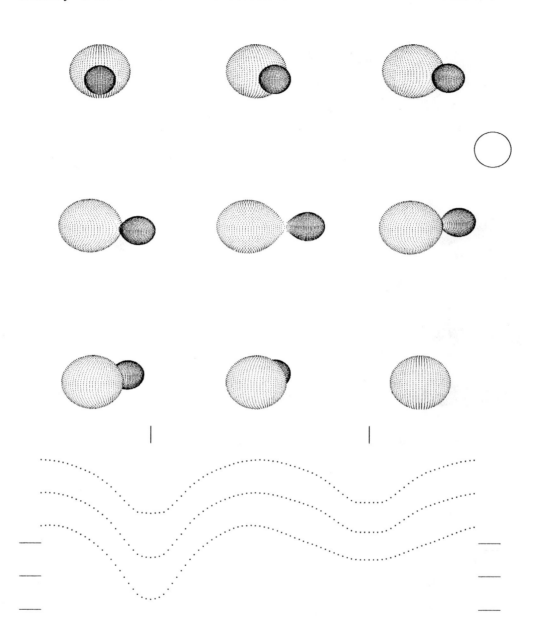

a= 3.23 $R_{\odot}$ r_1(pole)=0.472 r_2(pole)=0.253

e= 0.000 r_1(point)=−1.000 r_2(point)=−1.000

ω= − − − r_1(side)=0.509 r_2(side)=0.260

P=0^d.4215 r_1(back)=0.535 r_2(back)=0.298

i=$81°$.23 Ω_1=2.338 V_{γ}= −13 km sec^{-1}

T_1=6400 K Ω_2=2.338 F_1=1.0

T_2=5394 K   q=0.255 F_2=1.0

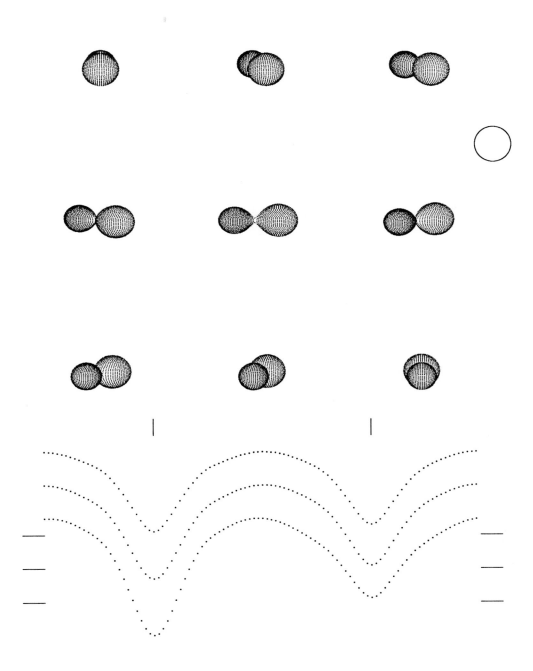

a=2.4 R$_\odot$ r$_1$(pole)=0.326   r$_2$(pole)=0.393

e= 0.000   r$_1$(point)=−1.000   r$_2$(point)=−1.000

ω= − − −   r$_1$(side)=0.341   r$_2$(side)=0.415

P=0^d.3117   r$_1$(back)=0.375   r$_2$(back)=0.446

i=82°.4 Ω_1=4.496   V$_\gamma$= − − −

T$_1$=5240 K Ω_2=4.496   F$_1$=1.00

T$_2$=4934 K   q=1.500   F$_2$=1.00

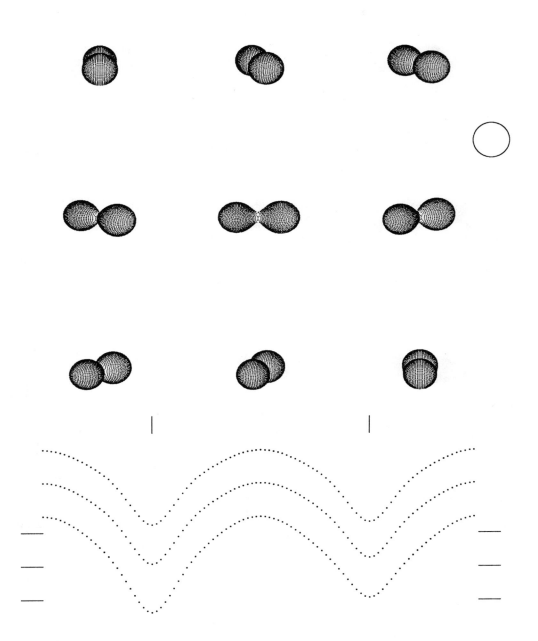

$$a=\ 2.42\ R_{\odot}$$

a= 2.42 R$_{\odot}$	r$_1$(pole)=0.359	r$_2$(pole)=0.381
e= 0.000	r$_1$(point)=−1.000	r$_2$(point)=−1.000
ω= −−−	r$_1$(side)=0.379	r$_2$(side)=0.402
P=0^d.3207	r$_1$(back)=0.418	r$_2$(back)=0.440
i=78°.95	Ω_1=3.851	V$_{\gamma}$= −22.5 km sec^{-1}
T$_1$=5630 K	Ω_2=3.851	F$_1$=1.00
T$_2$=5461 K	q=1.137	F$_2$=1.00

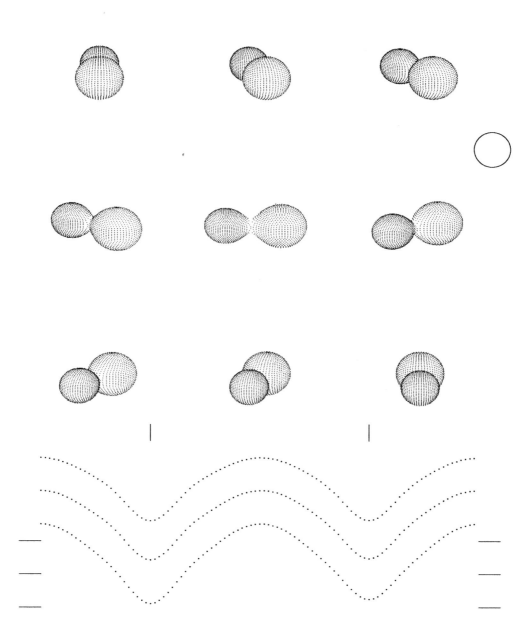

a= 3.0 $R_\odot$	r_1(pole)=0.335	r_2(pole)=0.411
e= 0.000	r_1(point)=−1.000	r_2(point)=−1.000
ω= −−−	r_1(side)=0.353	r_2(side)=0.438
P=0^d.3891	r_1(back)=0.396	r_2(back)=0.475
i=74°.42	Ω_1=4.493	V_γ= −−−
T_1=5500 K	Ω_2=4493.	F_1=1.00
T_2=5450 K	q=1.59	F_2=1.00

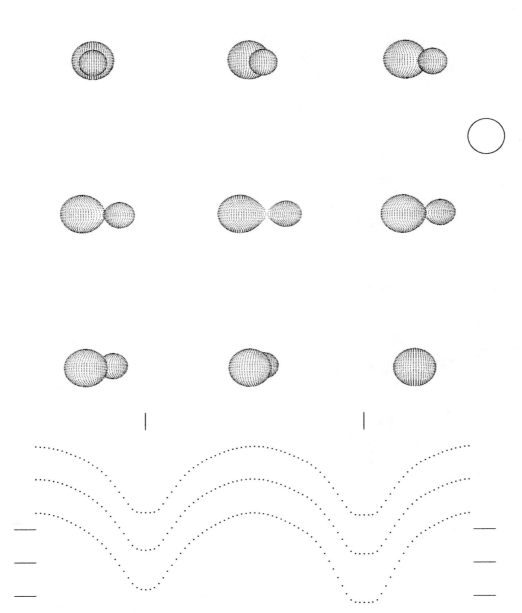

a=2.5 $R_\odot$ r_1(pole)=0.435 r_2(pole)=0.295

e= 0.000 r_1(point)=−1.000 r_2(point)=−1.000

ω= ––– r_1(side)=0.465 r_2(side)=0.308

P=0^d.3658 r_1(back)=0.495 r_2(back)=0.346

i=87°.0 Ω_1=2.691 V_γ= –––

T_1=6200 K Ω_2=2.691 F_1=1.00

T_2=6380 K   q=0.426 F_2=1.00

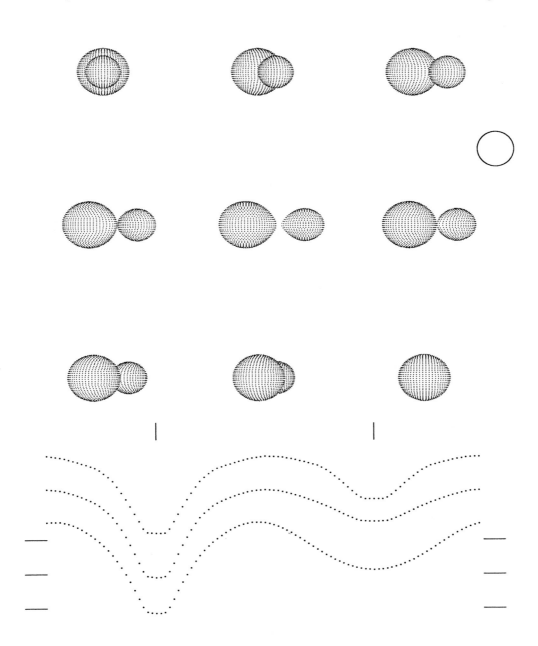

a=3.22 R$_\odot$ r$_1$(pole)=0.416   r$_2$(pole)=0.288

e= 0.000   r$_1$(point)=0.514   r$_2$(point)=0.414

ω= ---   r$_1$(side)=0.441   r$_2$(side)=0.300

P=0^d.4701   r$_1$(back)=0.465   r$_2$(back)=0.333

i=89°.7 Ω_1=2.798   V$_\gamma$=−9.6 km sec^{-1}

T$_1$=6700 K Ω_2=2.739   F$_1$=1.00

T$_2$=4020 K   q=0.430   F$_2$=1.00

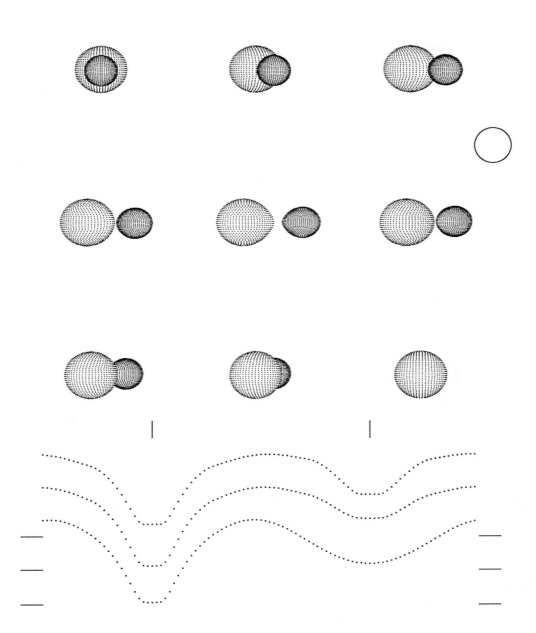

$a = 3.22\ R_\odot$ $r_1(\text{pole}) = 0.419$ $r_2(\text{pole}) = 0.274$

$e = 0.000$ $r_1(\text{point}) = 0.513$ $r_2(\text{point}) = 0.344$

$\omega = \text{---}$ $r_1(\text{side}) = 0.444$ $r_2(\text{side}) = 0.285$

$P = 0^d.4685$ $r_1(\text{back}) = 0.467$ $r_2(\text{back}) = 0.311$

$i = 88^\circ.5$ $\Omega_1 = 2.766$ $V_\gamma = -44.0\ \text{km sec}^{-1}$

$T_1 = 6800\ K$ $\Omega_2 = 2.753$ $F_1 = 1.00$

$T_2 = 4160\ K$ $q = 0.410$ $F_2 = 1.00$

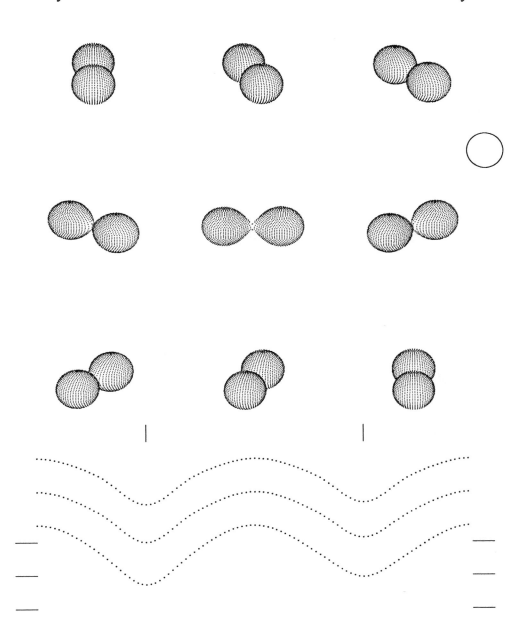

$$a=3.1\ R_\odot \qquad r_1(\text{pole})=0.356 \qquad r_2(\text{pole})=0.364$$

$$e=\ 0.000 \qquad r_1(\text{point})=-1.000 \qquad r_2(\text{point})=-1.000$$

$$\omega=\ --- \qquad r_1(\text{side})=0.374 \qquad r_2(\text{side})=0.384$$

$$P=0^d.4149 \qquad r_1(\text{back})=0.407 \qquad r_2(\text{back})=0.416$$

$$i=66°.6 \qquad \Omega_1=3.803 \qquad V_\gamma=\ ---$$

$$T_1=6709\ K \qquad \Omega_2=3.803 \qquad F_1=1.00$$

$$T_2=6371\ K \qquad q=1.053 \qquad F_2=1.00$$

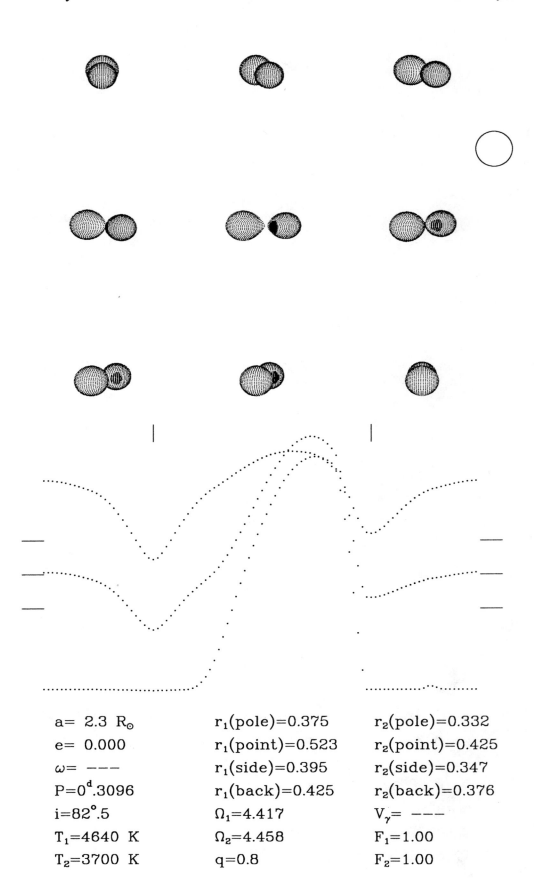

a= 2.3 $R_\odot$	r_1(pole)=0.375	r_2(pole)=0.332
e= 0.000	r_1(point)=0.523	r_2(point)=0.425
ω= ---	r_1(side)=0.395	r_2(side)=0.347
P=$0^d.3096$	r_1(back)=0.425	r_2(back)=0.376
i=$82°.5$	Ω_1=4.417	V_γ= ---
T_1=4640 K	Ω_2=4.458	F_1=1.00
T_2=3700 K	q=0.8	F_2=1.00

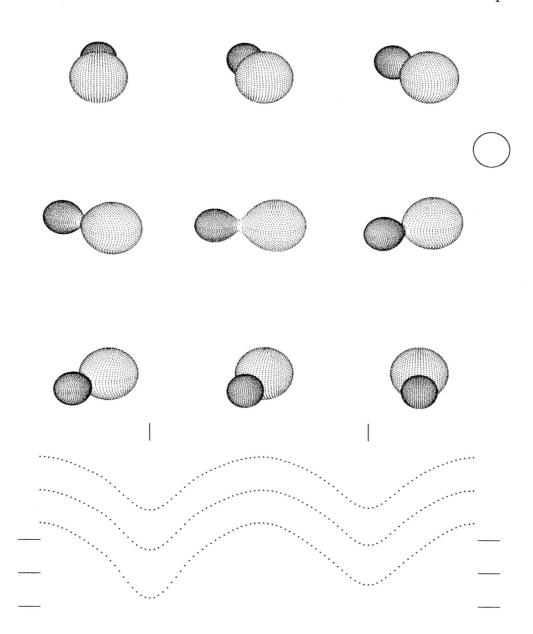

a= 3.3 $R_\odot$ r_1(pole)=0.286 r_2(pole)=0.445

e= 0.000 r_1(point)=−1.000 r_2(point)=−1.000

ω= – – – r_1(side)=0.299 r_2(side)=0.477

P=0^d.4534 r_1(back)=0.338 r_2(back)=0.507

i=$71°$.67 Ω_1=6.010 V_γ= – – –

T_1=6200 K Ω_2=6.010 F_1=1.00

T_2=5954 K   q=2.65 F_2=1.00

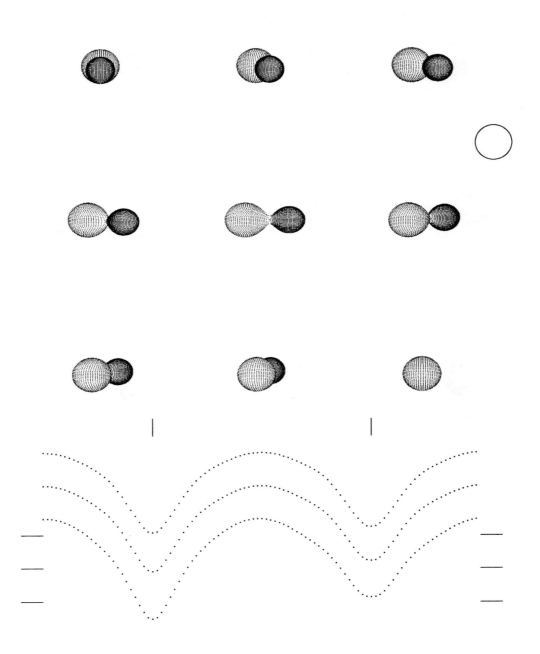

$a= 2.389\ R_{\odot}$ $r_1(pole)=0.412$ $r_2(pole)=0.319$

$e= 0.000$ $r_1(point)=-1.000$ $r_2(point)=-1.000$

$\omega=$ — — — $r_1(side)=0.439$ $r_2(side)=0.334$

$P=0^d.3448$ $r_1(back)=0.471$ $r_2(back)=0.372$

$i=84°.82$ $\Omega_1=2.949$ $V_{\gamma}=-39.84\ km\ sec^{-1}$

$T_1=6000\ K$ $\Omega_2=2.949$ $F_1=1.0$

$T_2=5800\ K$ $q=0.567$ $F_2=1.0$

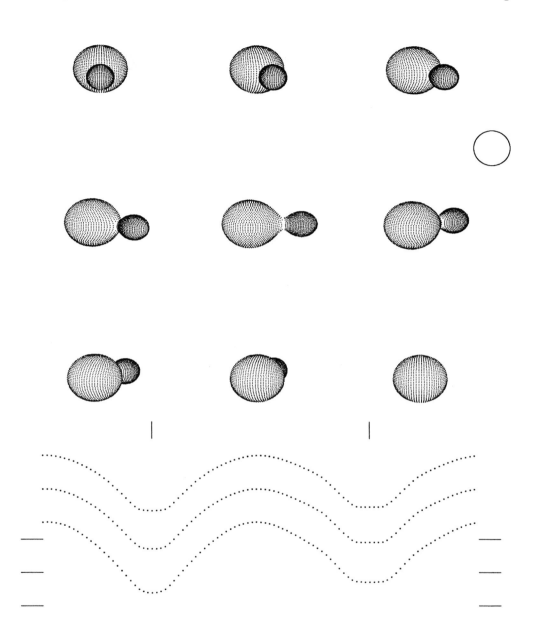

$a = 2.79\ R_{\odot}$ $r_1(\text{pole}) = 0.489$ $r_2(\text{pole}) = 0.261$

$e = 0.000$ $r_1(\text{point}) = -1.000$ $r_2(\text{point}) = -1.000$

$\omega = ---$ $r_1(\text{side}) = 0.534$ $r_2(\text{side}) = 0.274$

$P = 0^{d}.4096$ $r_1(\text{back}) = 0.563$ $r_2(\text{back}) = 0.324$

$i = 80°.30$ $\Omega_1 = 2.258$ $V_{\gamma} = -47.4\ \text{km sec}^{-1}$

$T_1 = 7000\ K$ $\Omega_2 = 2.258$ $F_1 = 1.0$

$T_2 = 6881\ K$ $q = 0.237$ $F_2 = 1.0$

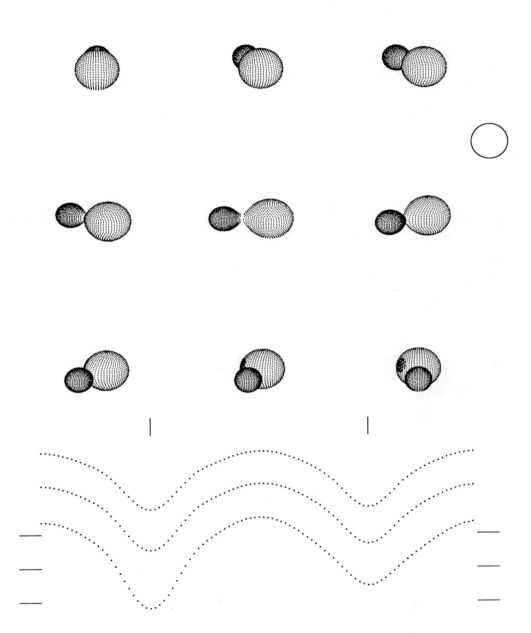

a= 2.532 R$_\odot$ r$_1$(pole)=0.282   r$_2$(pole)=0.450

e= 0.000   r$_1$(point)=−1.000   r$_2$(point)=−1.000

ω= − − −   r$_1$(side)=0.295   r$_2$(side)=0.483

P=0^d.3749   r$_1$(back)=0.334   r$_2$(back)=0.512

i=75°.41 Ω_1=6.242   V$_\gamma$=−29.85 km sec^{-1}

T$_1$=5800 K Ω_2=6.242   F$_1$=1.0

T$_2$=5522 K   q=2.804   F$_2$=1.0

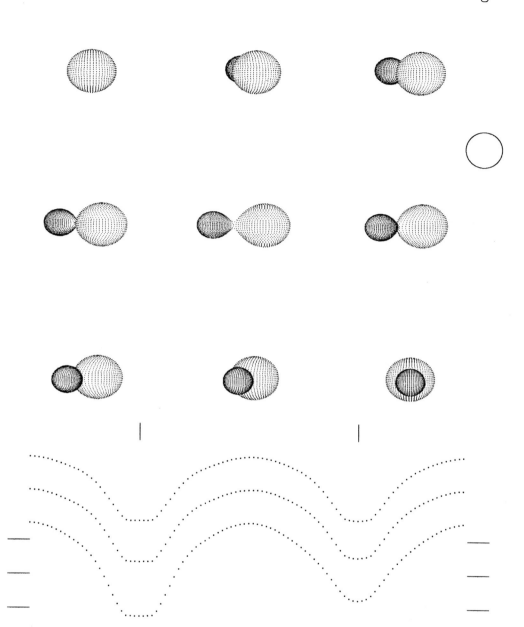

a= 2.8 R$_\odot$ r$_1$(pole)=0.278   r$_2$(pole)=0.447

e= 0.000   r$_1$(point)=−1.000   r$_2$(point)=−1.000

ω= ---   r$_1$(side)=0.290   r$_2$(side)=0.479

P=0^d.3615   r$_1$(back)=0.326   r$_2$(back)=0.507

i=86°.7 Ω_1=6.298   V$_\gamma$= ---

T$_1$=6545 K Ω_2=6.298   F$_1$=1.0

T$_2$=6210 K   q=2.811   F$_2$=1.0

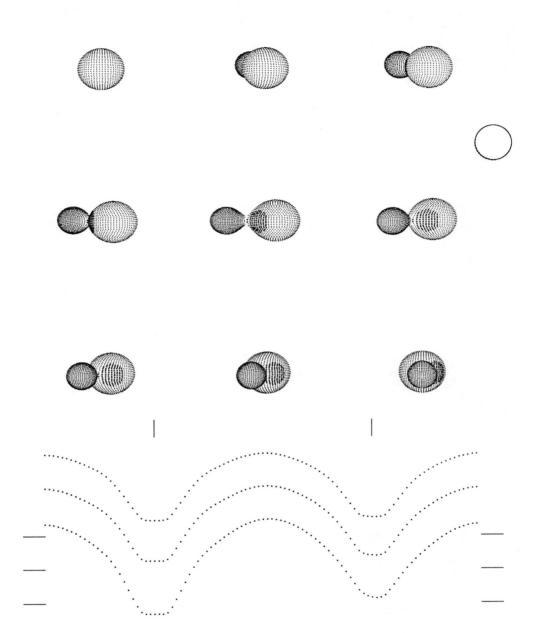

$$a= 2.68\ R_\odot \qquad r_1(\text{pole})=0.443 \qquad r_2(\text{pole})=0.289$$

$$e= 0.000 \qquad r_1(\text{point})=-1.000 \qquad r_2(\text{point})=-1.000$$

$$\omega= \text{---} \qquad r_1(\text{side})=0.475 \qquad r_2(\text{side})=0.303$$

$$P=0^d.3624 \qquad r_1(\text{back})=0.505 \qquad r_2(\text{back})=0.342$$

$$i=86^\circ.26 \qquad \Omega_1=2.613 \qquad V_\gamma=-11.3\ \text{km sec}^{-1}$$

$$T_1=6000\ K \qquad \Omega_2=2.613 \qquad F_1=1.00$$

$$T_2=6164\ K \qquad q=0.39 \qquad F_2=1.00$$

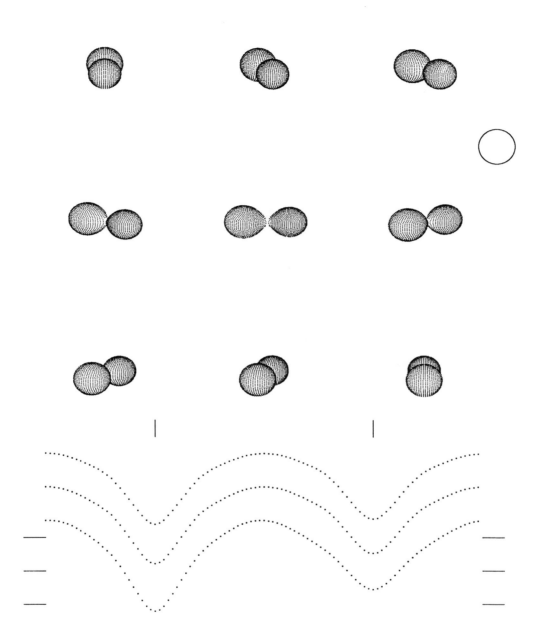

a= 2.5 R$_\odot$ r$_1$(pole)=0.378   r$_2$(pole)=0.344

e= 0.000   r$_1$(point)=−1.000   r$_2$(point)=−1.000

ω= −−−   r$_1$(side)=0.400   r$_2$(side)=0.361

P=0^d.3605   r$_1$(back)=0.432   r$_2$(back)=0.396

i=77°.45 Ω_1=3.405   V$_\gamma$= −−−

T$_1$=5600 K Ω_2=3.405   F$_1$=1.0

T$_2$=5325 K   q=0.813   F$_2$=1.0

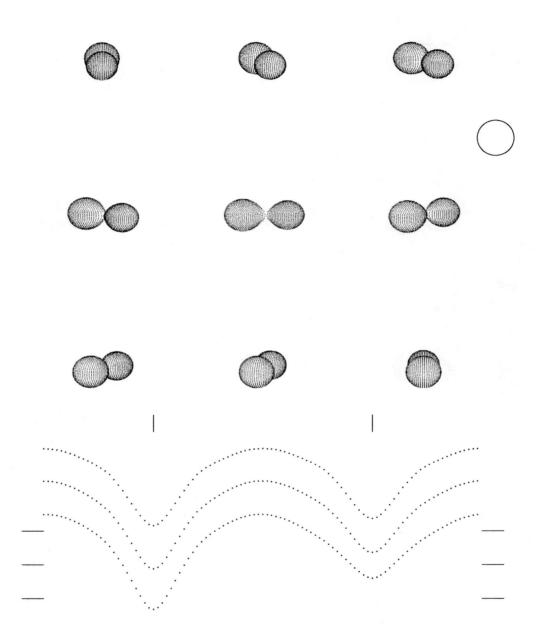

a= 2.4 $R_\odot$ r_1(pole)=0.380 r_2(pole)=0.343

e= 0.000 r_1(point)=−1.000 r_2(point)=−1.000

ω= --- r_1(side)=0.402 r_2(side)=0.360

P=0^d.3865 r_1(back)=0.434 r_2(back)=0.395

i=80°.2 Ω_1=3.378 V_γ= ---

T_1=5100 K Ω_2=3.378 F_1=1.0

T_2=4780 K   q=0.80 F_2=1.0

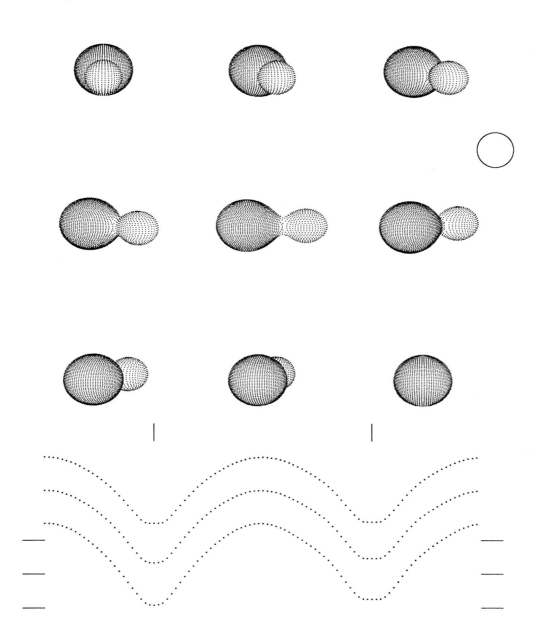

a= 3.19 R$_\odot$	r$_1$(pole)=0.463	r$_2$(pole)=0.304
e= 0.000	r$_1$(point)=−1.000	r$_2$(point)=−1.000
ω= −−−	r$_1$(side)=0.502	r$_2$(side)=0.321
P=0^d.416	r$_1$(back)=0.539	r$_2$(back)=0.381
i=82°.88	Ω_1=2.496	V$_\gamma$=+5.0 km sec^{-1}
T$_1$=7200 K	Ω_2=2.496	F$_1$=1.00
T$_2$=7146 K	q=0.372	F$_2$=1.00

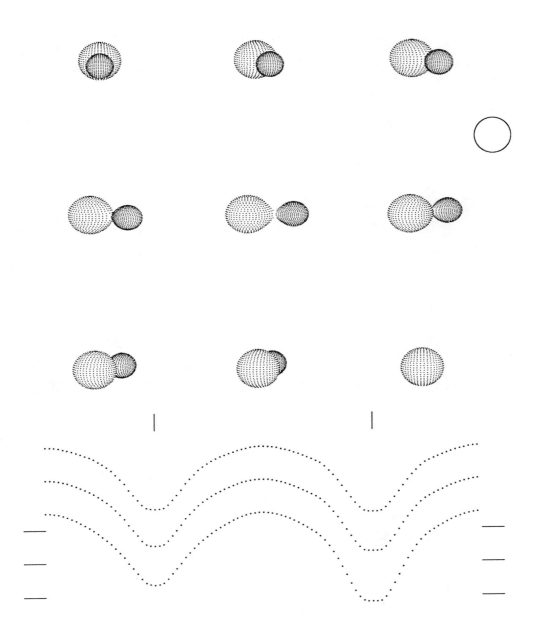

$a=2.5\ R_\odot$ $r_1(pole)=0.438$ $r_2(pole)=0.288$

$e=0.000$ $r_1(point)=-1.000$ $r_2(point)=-1.000$

$\omega=$ ——— $r_1(side)=0.469$ $r_2(side)=0.301$

$P=0^d.3544$ $r_1(back)=0.498$ $r_2(back)=0.338$

$i=83°.3$ $\Omega_1=2.648$ $V_\gamma=$ ———

$T_1=5500\ K$ $\Omega_2=2.648$ $F_1=1.00$

$T_2=5689\ K$ $q=0.400$ $F_2=1.00$

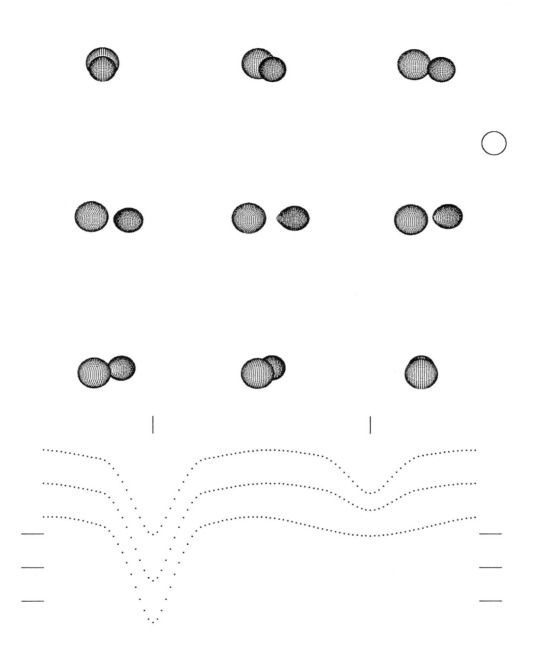

a= 2.79 R$_\odot$ r$_1$(pole)=0.349   r$_2$(pole)=0.282

e= 0.000   r$_1$(point)=0.377   r$_2$(point)=0.408

ω= ---   r$_1$(side)=0.360   r$_2$(side)=0.295

P=0^d.6874   r$_1$(back)=0.370   r$_2$(back)=0.328

i=82°.3 Ω_1=3.247   V$_\gamma$=−13.0 km sec^{-1}

T$_1$=9750 K Ω_2=2.682   F$_1$=1.00

T$_2$=5600 K   q=0.402   F$_2$=1.00

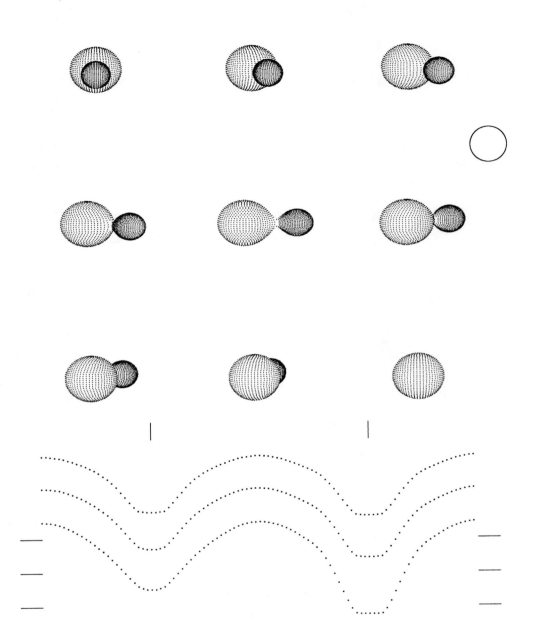

a=2.87 R$_\odot$ r$_1$(pole)=0.458   r$_2$(pole)=0.272

e= 0.000   r$_1$(point)=−1.000   r$_2$(point)=−1.000

ω= ---   r$_1$(side)=0.493   r$_2$(side)=0.284

P=0^d.4075   r$_1$(back)=0.521   r$_2$(back)=0.322

i=84°.1 Ω_1=2.474   V$_\gamma$=10.0 km sec^{-1}

T$_1$=5010 K Ω_2=2.474   F$_1$=1.00

T$_2$=5275 K   q=0.318   F$_2$=1.00

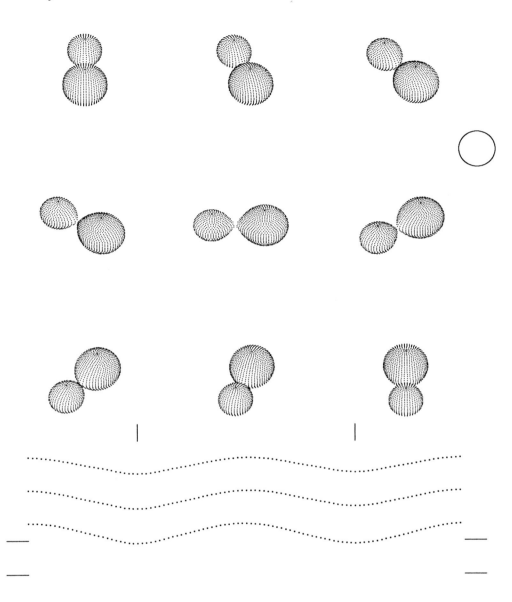

$a=2.9\ R_\odot$ $r_1(\text{pole})=0.312$ $r_2(\text{pole})=0.402$

$e=0.000$ $r_1(\text{point})=0.429$ $r_2(\text{point})=0.540$

$\omega=\ ---$ $r_1(\text{side})=0.326$ $r_2(\text{side})=0.425$

$P=0^d.3638$ $r_1(\text{back})=0.358$ $r_2(\text{back})=0.454$

$i=47^\circ.62$ $\Omega_1=4.856$ $V_\gamma=\ ---$

$T_1=5780\ K$ $\Omega_2=4.856$ $F_1=1.00$

$T_2=5487\ K$ $q=1.720$ $F_2=1.00$

Group III

Systems with
$3.38\,R_\odot < a \leqslant 5.06\,R_\odot$

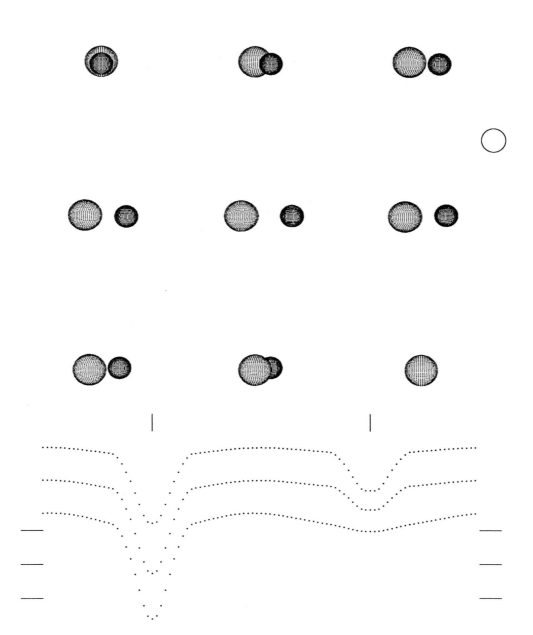

a= 4.18 $R_\odot$ r_1(pole)=0.315 r_2(pole)=0.224

e= 0.000 r_1(point)=0.341 r_2(point)=0.235

ω= ——— r_1(side)=0.324 r_2(side)=0.227

P=0^d.6289 r_1(back)=0.334 r_2(back)=0.233

i=87°.30 Ω_1=3.791 V_γ= +20 km sec^{-1}

T_1=6250 K Ω_2=4.058 F_1=1.00

T_2=4813 K   q=0.65 F_2=1.00

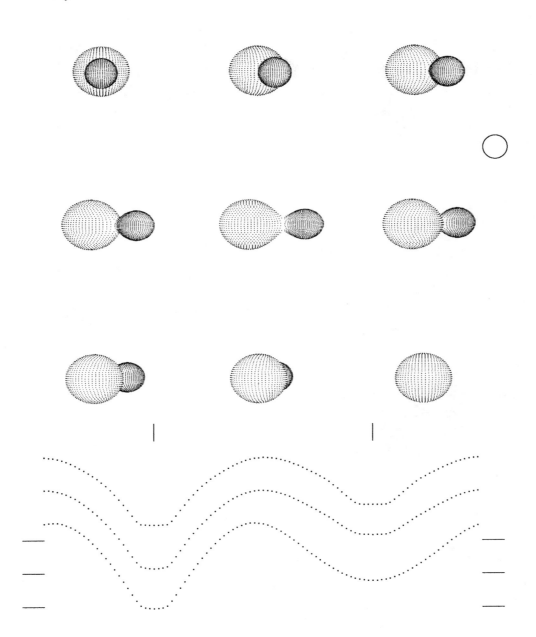

a= 4.5 R$_\odot$ r$_1$(pole)=0.473   r$_2$(pole)=0.286

e= 0.000   r$_1$(point)=−1.000   r$_2$(point)=−1.000

ω= −−−   r$_1$(side)=0.514   r$_2$(side)=0.301

P=0^d.7224   r$_1$(back)=0.548   r$_2$(back)=0.356

i=88°.0 Ω_1=2.394   V$_\gamma$= −−−

T$_1$=7500 K Ω_2=2.394   F$_1$=1.0

T$_2$=5370 K   q=0.311   F$_2$=1.0

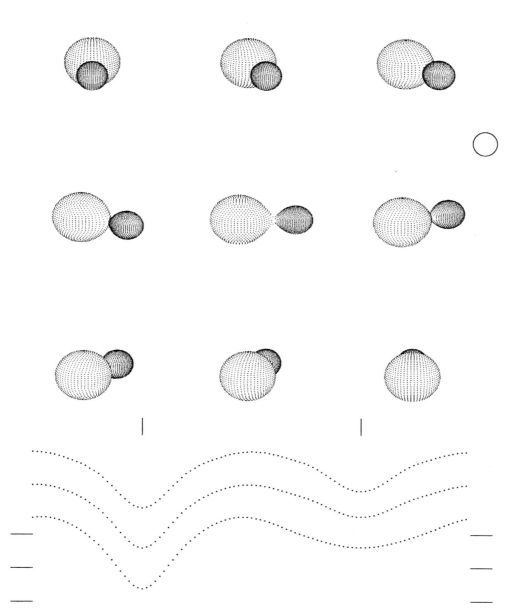

a= 4.6 R$_\odot$ r$_1$(pole)=0.460   r$_2$(pole)=0.266

e= 0.000   r$_1$(point)=−1.000   r$_2$(point)=−1.000

ω= −−−   r$_1$(side)=0.496   r$_2$(side)=0.278

P=0^d.7810   r$_1$(back)=0.523   r$_2$(back)=0.314

i=75°.5 Ω_1=2.444   V$_\gamma$= −−−

T$_1$=6500 K Ω_2=2.444   F$_1$=1.0

T$_2$=4890 K   q=0.3   F$_2$=1.0

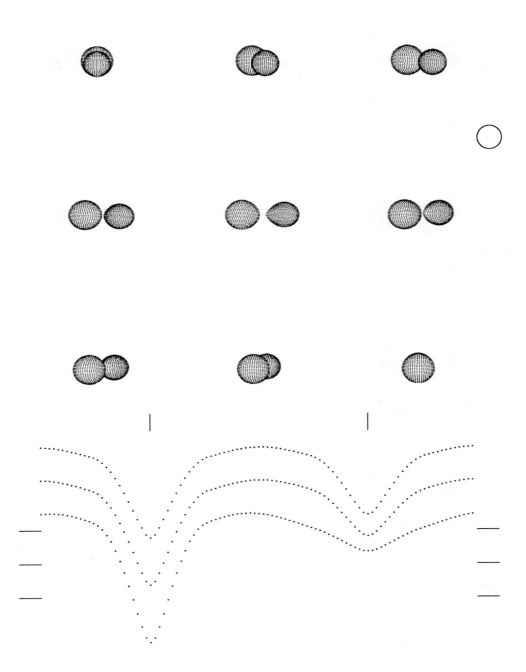

$$
\begin{array}{lll}
a= 3.47\ R_\odot & r_1(\text{pole})=0.365 & r_2(\text{pole})=0.313 \\
e= 0.000 & r_1(\text{point})=0.419 & r_2(\text{point})=0.446 \\
\omega= \text{---} & r_1(\text{side})=0.381 & r_2(\text{side})=0.327 \\
P=0^d.5560 & r_1(\text{back})=0.398 & r_2(\text{back})=0.359 \\
i=86°.2 & \Omega_1=3.294 & V_\gamma= +13\ \text{km sec}^{-1} \\
T_1=6400\ K & \Omega_2=3.045 & F_1=1.00 \\
T_2=5289\ K & q=0.59 & F_2=1.00
\end{array}
$$

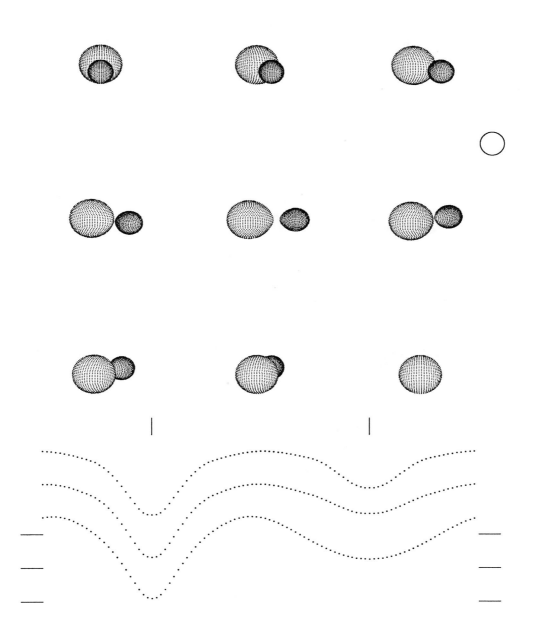

a= 3.82 R$_\odot$ r$_1$(pole)=0.434   r$_2$(pole)=0.262

e= 0.000   r$_1$(point)=0.522   r$_2$(point)=0.344

ω= – – –   r$_1$(side)=0.462   r$_2$(side)=0.272

P=0^d.5090   r$_1$(back)=0.482   r$_2$(back)=0.302

i=80°.5 Ω_1=2.598   V$_\gamma$=−2.3 km sec^{-1}

T$_1$=7060 K Ω_2=2.528   F$_1$=1.0

T$_2$=4395 K   q=0.32   F$_2$=1.0

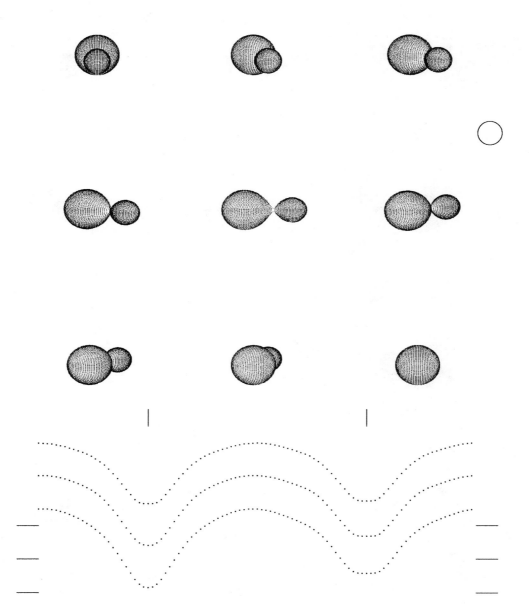

a= 3.86 $R_\odot$ r_1(pole)=0.361 r_2(pole)=0.276

e= 0.000 r_1(point)=−1.000 r_2(point)=−1.000

ω= ——— r_1(side)=0.442 r_2(side)=0.287

P=0^d.6293 r_1(back)=0.473 r_2(back)=0.321

i=$82°$.14 Ω_1=2.592 V_γ= −17.6 km sec^{-1}

T_1=8750 K Ω_2=2.592 F_1=1.0

T_2=8572 K   q=0.361 F_2=1.0

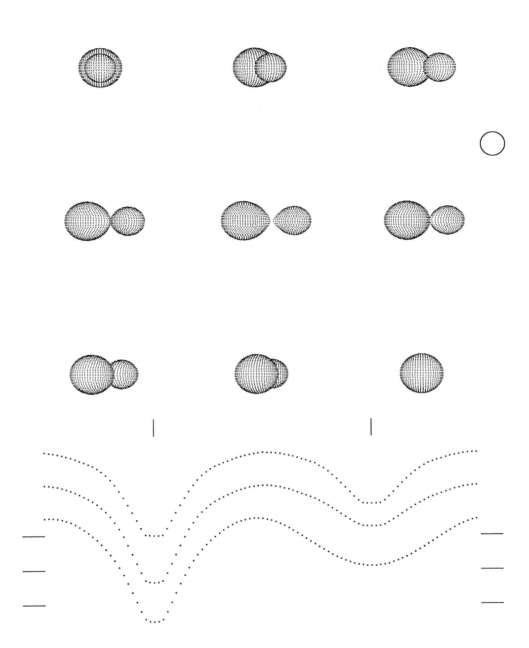

a=4.1 R$_{\odot}$	r$_1$(pole)=0.409	r$_2$(pole)=0.301
e= 0.000	r$_1$(point)=0.525	r$_2$(point)=0.429
ω= – – –	r$_1$(side)=0.433	r$_2$(side)=0.315
P=0$^{\rm d}$.6012	r$_1$(back)=0.460	r$_2$(back)=0.348
i=89°.7	Ω_1=2.920	V$_{\gamma}$= – – –
T$_1$=8200 K	Ω_2=2.897	F$_1$=1.00
T$_2$=5405 K	q=0.511	F$_2$=1.00

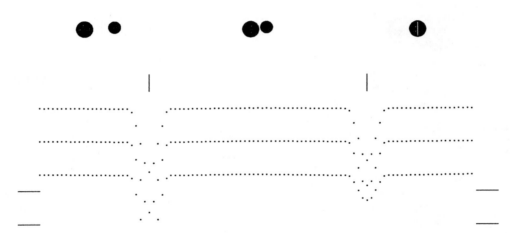

a=4.2 $R_\odot$ r_1(pole)=0.161 r_2(pole)=0.120

e= 0.000 r_1(point)=0.163 r_2(point)=0.121

ω= ——— r_1(side)=0.162 r_2(side)=0.120

P=0^d.7104 r_1(back)=0.163 r_2(back)=0.121

i=87°.4 Ω_1=6.986 V_γ= ———

T_1=5900 K Ω_2=7.767 F_1=1.00

T_2=5390 K   q=0.800 F_2=1.00

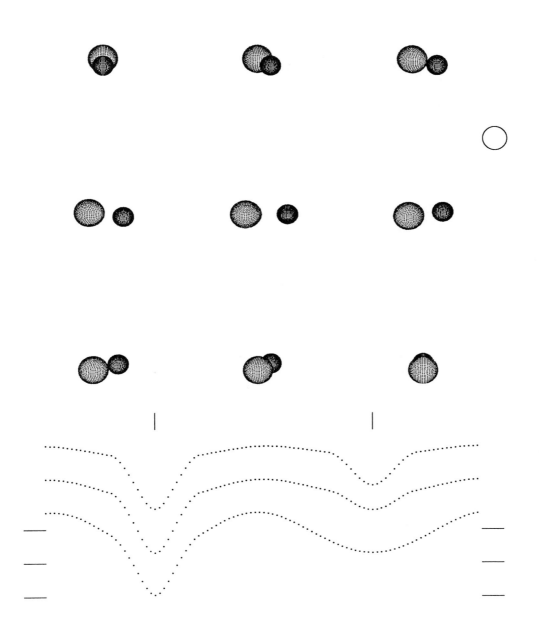

$$a=\ 3.46\ R_{\odot} \qquad r_1(pole)=0.340 \qquad r_2(pole)=0.240$$

$$e=\ 0.000 \qquad r_1(point)=0.382 \qquad r_2(point)=0.255$$

$$\omega=\ --- \qquad r_1(side)=0.352 \qquad r_2(side)=0.244$$

$$P=0^d.5931 \qquad r_1(back)=0.367 \qquad r_2(back)=0.251$$

$$i=80^\circ.0 \qquad \Omega_1=3.603 \qquad V_\gamma=-15.0\ km\ sec^{-1}$$

$$T_1=5800\ K \qquad \Omega_2=4.041 \qquad F_1=1.00$$

$$T_2=4140\ K \qquad q=0.700 \qquad F_2=1.00$$

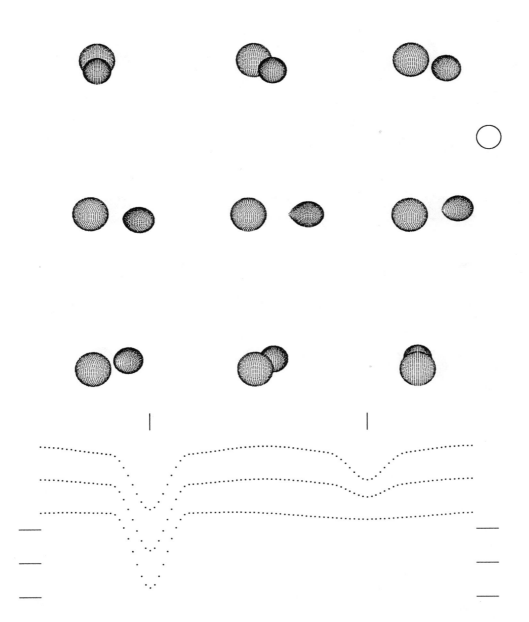

a= 4.92 R$_\odot$ r$_1$(pole)=0.294   r$_2$(pole)=0.226

e= 0.000   r$_1$(point)=0.302   r$_2$(point)=0.332

ω= ———   r$_1$(side)=0.299   r$_2$(side)=0.235

P=1^d.1359   r$_1$(back)=0.301   r$_2$(back)=0.267

i=79°.4 Ω_1=3.570   V$_\gamma$=−32.7 km sec^{-1}

T$_1$=7220 K Ω_2=2.182   F$_1$=1.00

T$_2$=4650 K   q=0.180   F$_2$=1.00

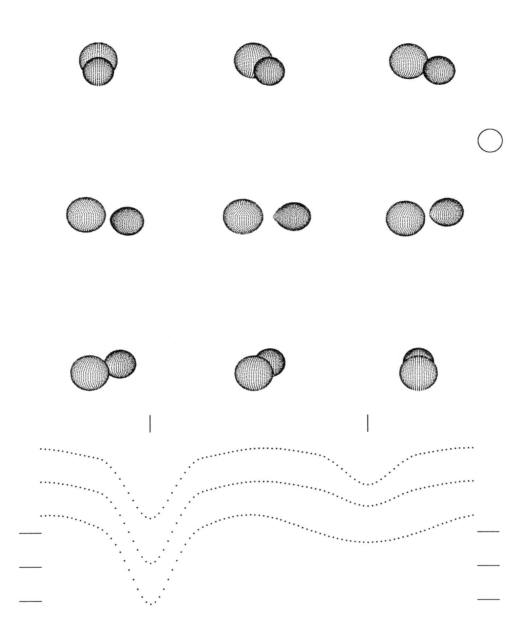

$a= 4.2\ R_\odot$ $r_1(\text{pole})=0.360$ $r_2(\text{pole})=0.290$

$e= 0.000$ $r_1(\text{point})=0.396$ $r_2(\text{point})=0.416$

$\omega= \text{---}$ $r_1(\text{side})=0.374$ $r_2(\text{side})=0.302$

$P=0^d.5788$ $r_1(\text{back})=0.385$ $r_2(\text{back})=0.335$

$i=77°.09$ $\Omega_1=3.188$ $V_\gamma= \text{---}$

$T_1=8200\ K$ $\Omega_2=2.755$ $F_1=1.00$

$T_2=4837\ K$ $q=0.438$ $F_2=1.00$

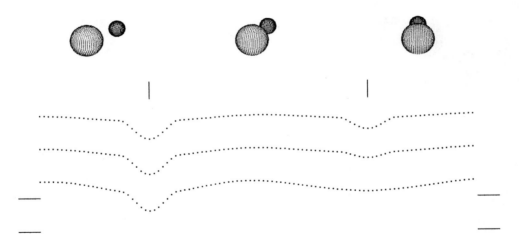

a= 4.23 R$_\odot$	r$_1$(pole)=0.312	r$_2$(pole)=0.159
e= 0.000	r$_1$(point)=0.333	r$_2$(point)=0.162
ω= ---	r$_1$(side)=0.320	r$_2$(side)=0.160
P=1^d.0228	r$_1$(back)=0.328	r$_2$(back)=0.162
i=73°.0	Ω_1=3.728	V$_\gamma$=−0.2 km sec^{-1}
T$_1$=7000 K	Ω_2=4.665	F$_1$=1.0
T$_2$=4900 K	q=0.55	F$_2$=1.0

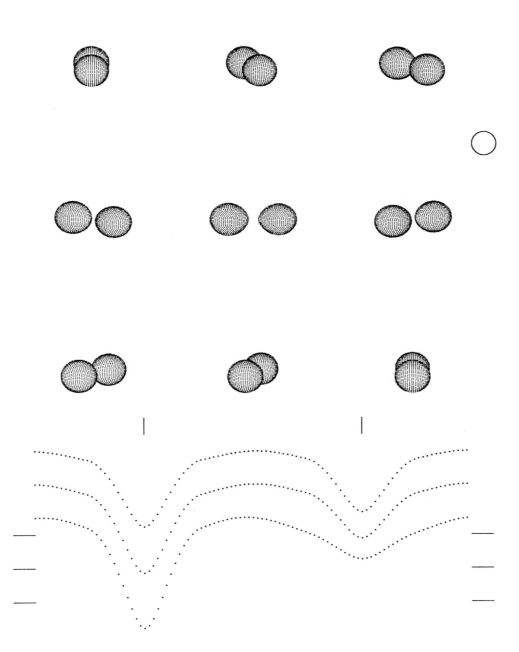

a=4.1 R$_\odot$ r$_1$(pole)=0.342   r$_2$(pole)=0.331

e= 0.000   r$_1$(point)=0.401   r$_2$(point)=0.404

ω= — — —   r$_1$(side)=0.356   r$_2$(side)=0.345

P=0^d.7591   r$_1$(back)=0.377   r$_2$(back)=0.370

i=80°.4 Ω_1=3.748   V$_\gamma$= — — —

T$_1$=5530 K Ω_2=3.645   F$_1$=1.00

T$_2$=4780 K   q=0.871   F$_2$=1.00

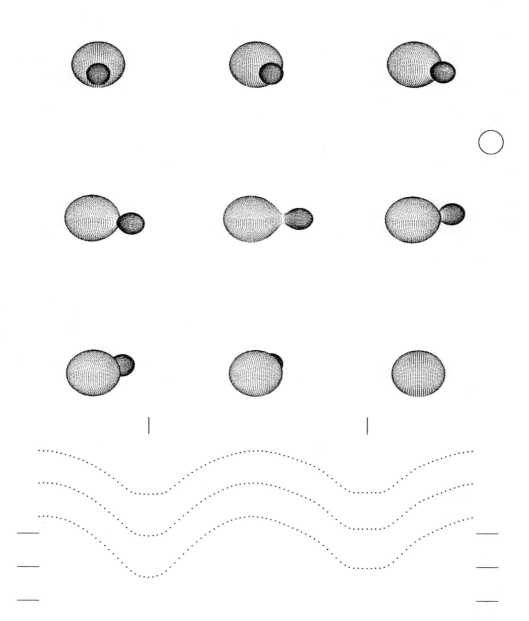

$a= 3.94\ R_\odot$ $r_1(\text{pole})=0.506$ $r_2(\text{pole})=0.237$

$e= 0.000$ $r_1(\text{point})=-1.000$ $r_2(\text{point})=-1.000$

$\omega= \text{---}$ $r_1(\text{side})=0.556$ $r_2(\text{side})=0.247$

$P=0^d.6057$ $r_1(\text{back})=0.581$ $r_2(\text{back})=0.292$

$i=78^\circ.10$ $\Omega_1=2.132$ $V_\gamma= -16\ \text{km sec}^{-1}$

$T_1=7250\ K$ $\Omega_2=2.132$ $F_1=1.0$

$T_2=7158\ K$ $q=0.178$ $F_2=1.0$

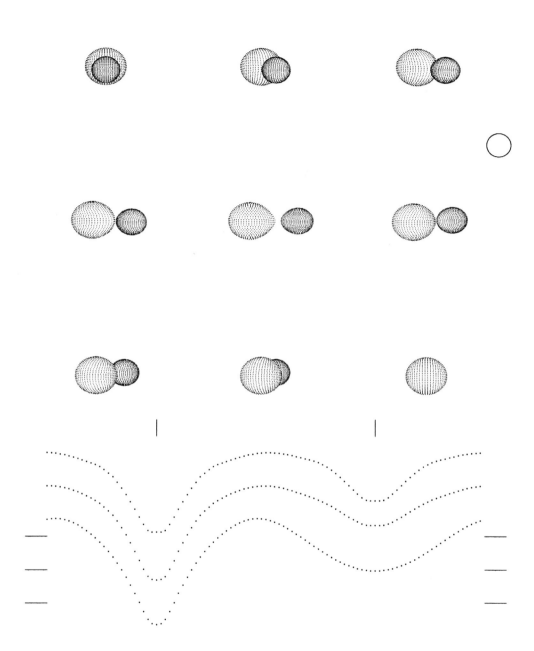

a= 3.9 R$_\odot$ r$_1$(pole)=0.409   r$_2$(pole)=0.290

e= 0.000   r$_1$(point)=0.521   r$_2$(point)=0.362

ω= – – –   r$_1$(side)=0.434   r$_2$(side)=0.301

P=0^d.5446   r$_1$(back)=0.460   r$_2$(back)=0.328

i=85°.7 Ω_1=2.906   V$_\gamma$= – – –

T$_1$=7800 K Ω_2=2.935   F$_1$=1.00

T$_2$=4930 K   q=0.50   F$_2$=1.00

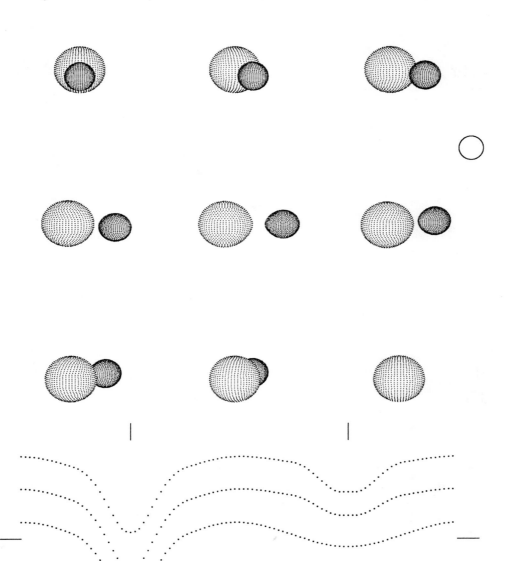

a=4.8 R$_\odot$	r$_1$(pole)=0.414	r$_2$(pole)=0.253
e= 0.000	r$_1$(point)=0.478	r$_2$(point)=0.308
ω= ---	r$_1$(side)=0.437	r$_2$(side)=0.262
P=0^d.7408	r$_1$(back)=0.454	r$_2$(back)=0.286
i=83°.0	Ω_1=2.718	V$_\gamma$= ---
T$_1$=9780 K	Ω_2=2.608	F$_1$=1.00
T$_2$=5700 K	q=0.33	F$_2$=1.00

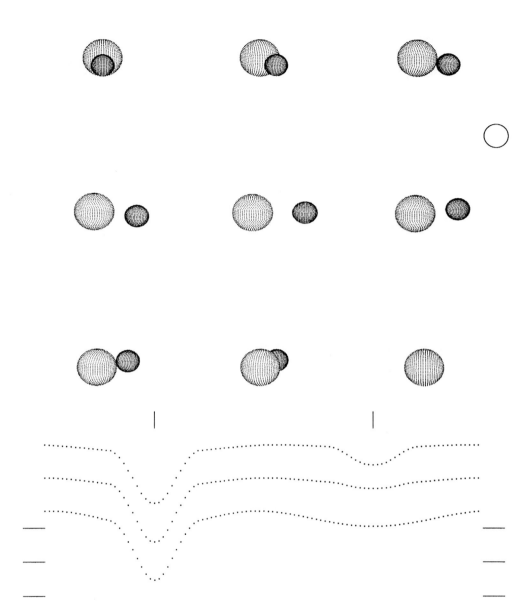

a= 4.31 $R_\odot$ r_1(pole)=0.364 r_2(pole)=0.214

e= 0.000 r_1(point)=0.388 r_2(point)=0.242

ω= --- r_1(side)=0.375 r_2(side)=0.220

P=0^d.7264 r_1(back)=0.382 r_2(back)=0.234

i=$81°$.48 Ω_1=2.992 V_γ=-11 km sec^{-1}

T_1=7500 K Ω_2=2.542 F_1=1.00

T_2=3769 K   q=0.256 F_2=1.00

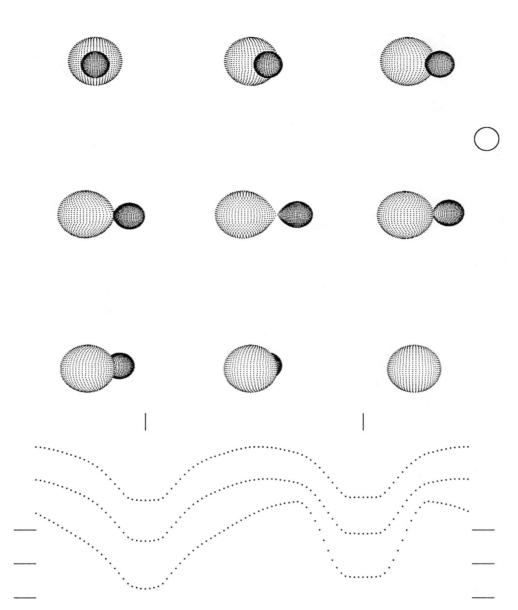

a=4.42 $R_\odot$ r_1(pole)=0.464 r_2(pole)=0.253

e= 0.000 r_1(point)=0.621 r_2(point)=0.359

ω= ——— r_1(side)=0.500 r_2(side)=0.264

P=$0^d.7473$ r_1(back)=0.525 r_2(back)=0.296

i=$86°.3$ Ω_1=2.399 V_γ=19.0 km sec^{-1}

T_1=6600 K Ω_2=2.399 F_1=1.00

T_2=5792 K   q=0.270 F_2=1.00

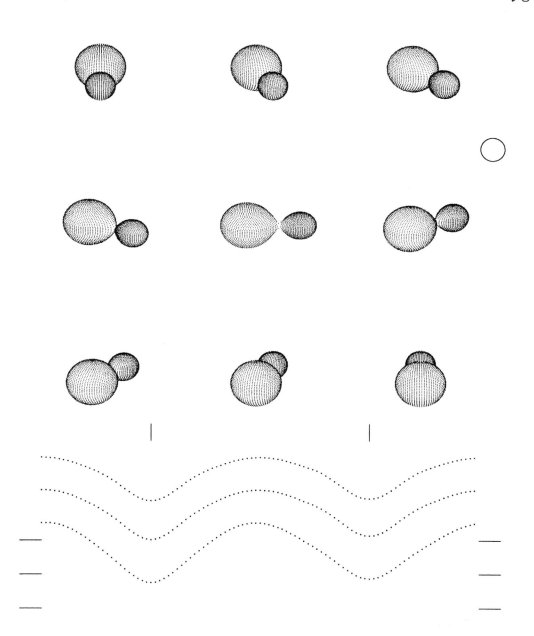

a= 4.33 $R_\odot$	r_1(pole)=0.449	r_2(pole)=0.276
e= 0.000	r_1(point)=−1.000	r_2(point)=−1.000
ω= −−−	r_1(side)=0.481	r_2(side)=0.285
P=0^d.7859	r_1(back)=0.508	r_2(back)=0.320
i=$68°$.3	Ω_1=2.538	V_γ= −8.1 km sec^{-1}
T_1=8750 K	Ω_2=2.538	F_1=1.00
T_2=8207 K	q=0.34	F_2=1.00

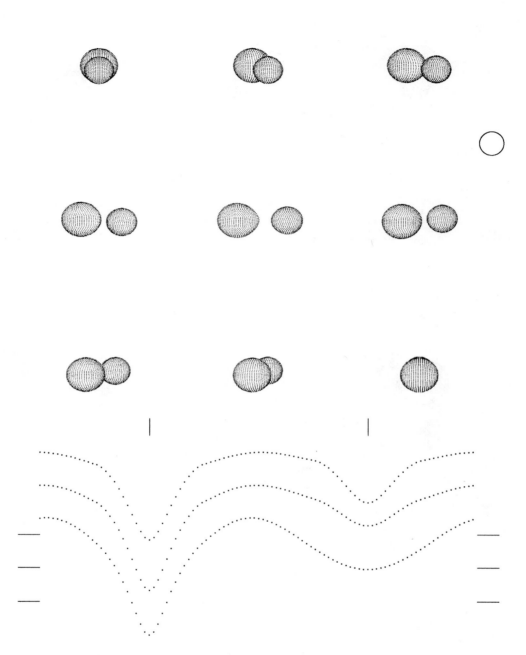

a= 4.1 $R_\odot$ r_1(pole)=0.362 r_2(pole)=0.289

e= 0.000 r_1(point)=0.437 r_2(point)=0.323

ω= ——— r_1(side)=0.378 r_2(side)=0.297

P=0^d.5682 r_1(back)=0.402 r_2(back)=0.312

i=84°.38 Ω_1=3.495 V_γ= ———

T_1=7380 K Ω_2=3.759 F_1=1.00

T_2=4748 K   q=0.775 F_2=1.00

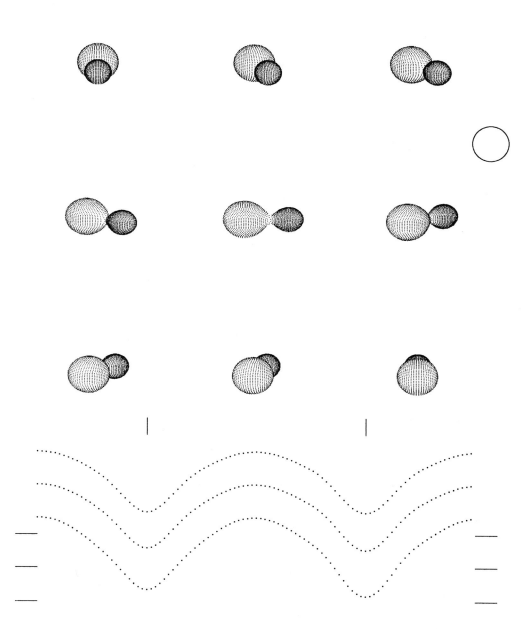

a= 2.38 $R_\odot$ r_1(pole)=0.441 r_2(pole)=0.295

e= 0.000 r_1(point)=−1.000 r_2(point)=−1.000

ω= − − − r_1(side)=0.473 r_2(side)=0.310

P=0^d.3501 r_1(back)=0.504 r_2(back)=0.350

i=76°.28 Ω_1=2.645 V_γ=−61.2 km sec^{-1}

T_1=6245 K Ω_2=2.645 F_1=1.0

T_2=6345 K   q=0.411 F_2=1.0

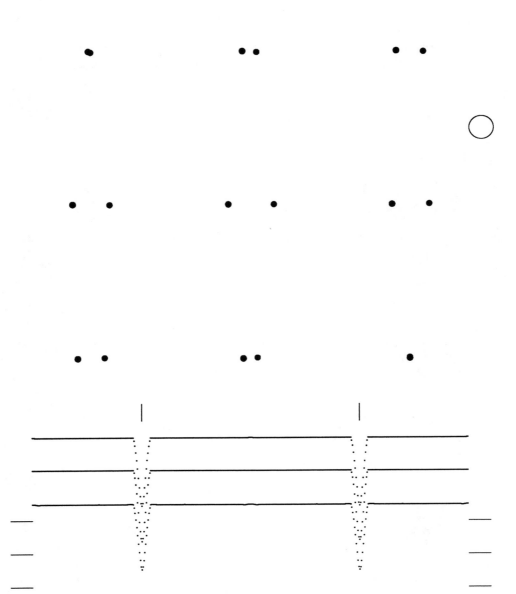

a= 3.76 $R_{\odot}$ r_1(pole)=0.067 r_2(pole)=0.062

e= 0.000 r_1(point)=0.067 r_2(point)=0.062

ω= ——— r_1(side)=0.067 r_2(side)=0.062

P=1^d.2684 r_1(back)=0.067 r_2(back)=0.062

i=89°.82 Ω_1=15.780 V_{γ}= −118 km sec^{-1}

T_1=3150 K Ω_2=15.008 F_1=1.00

T_2=3150 K   q=0.87 F_2=1.00

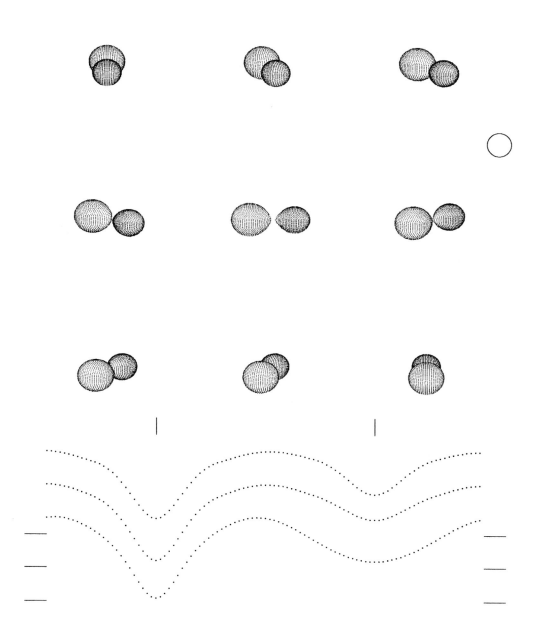

a=3.63 R$_\odot$ r$_1$(pole)=0.385   r$_2$(pole)=0.314

e= 0.000   r$_1$(point)=0.470   r$_2$(point)=0.448

ω= ---   r$_1$(side)=0.405   r$_2$(side)=0.328

P=0^d.6322   r$_1$(back)=0.429   r$_2$(back)=0.361

i=75°.1 Ω_1=3.154   V$_\gamma$=−20.1 km sec^{-1}

T$_1$=7300 K Ω_2=3.063   F$_1$=1.00

T$_2$=4650 K   q=0.600   F$_2$=1.00

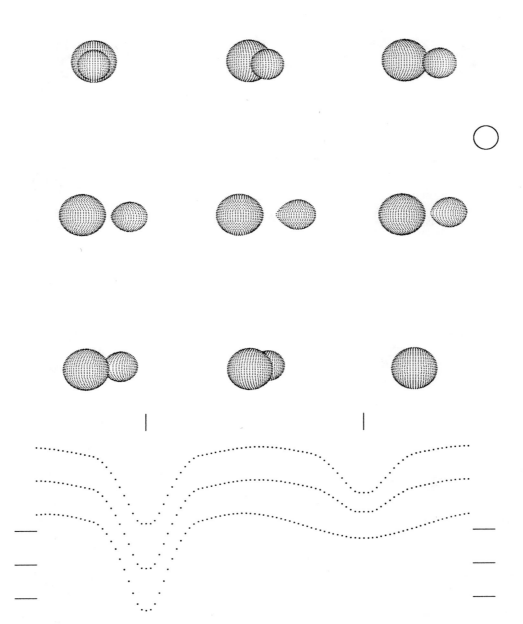

$a=4.8\ R_\odot$ $r_1(\text{pole})=0.376$ $r_2(\text{pole})=0.269$

$e=\ 0.000$ $r_1(\text{point})=0.409$ $r_2(\text{point})=0.389$

$\omega=\ ---$ $r_1(\text{side})=0.389$ $r_2(\text{side})=0.280$

$P=0^{d}.8233$ $r_1(\text{back})=0.399$ $r_2(\text{back})=0.313$

$i=86°.38$ $\Omega_1=2.983$ $V_\gamma=\ ---$

$T_1=7500\ K$ $\Omega_2=2.541$ $F_1=1.00$

$T_2=5095\ K$ $q=0.334$ $F_2=1.00$

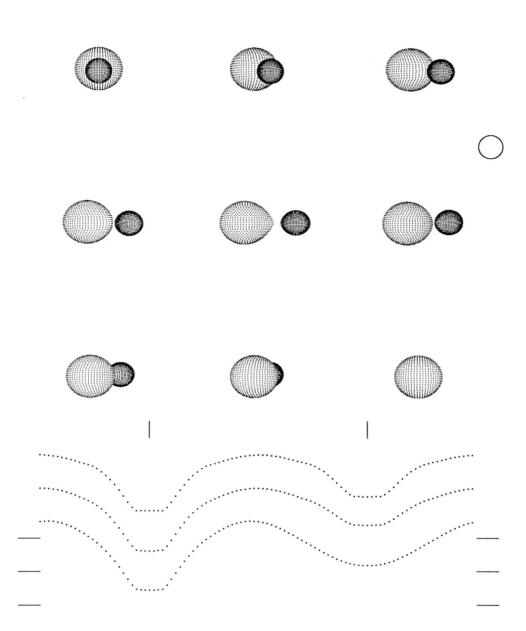

a=4.2 R$_\odot$ r$_1$(pole)=0.437   r$_2$(pole)=0.252

e= 0.000   r$_1$(point)=0.557   r$_2$(point)=0.296

ω= –––   r$_1$(side)=0.466   r$_2$(side)=0.260

P=0$^{\rm d}$.6385   r$_1$(back)=0.491   r$_2$(back)=0.280

i=87°.5 Ω_1=2.619   V$_\gamma$= –––

T$_1$=7580 K Ω_2=2.720   F$_1$=1.00

T$_2$=5490 K   q=0.360   F$_2$=1.00

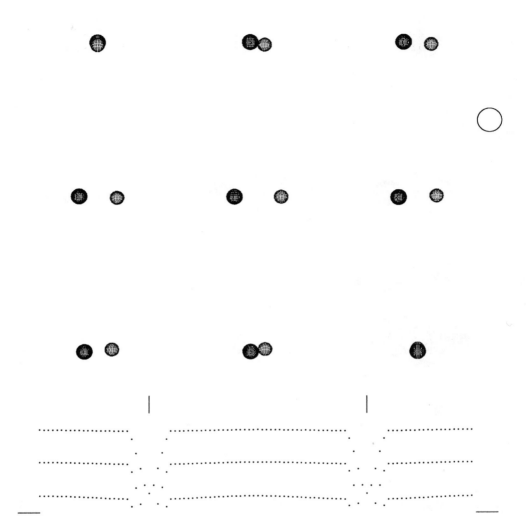

a= 3.85 R$_\odot$ r$_1$(pole)=0.170   r$_2$(pole)=0.149

e= 0.000   r$_1$(point)=0.172   r$_2$(point)=0.150

ω= ———   r$_1$(side)=0.171   r$_2$(side)=0.150

P=0^d.8143   r$_1$(back)=0.172   r$_2$(back)=0.150

i=86°.54 Ω_1=6.864   V$_\gamma$=+0.9 km sec^{-1}

T$_1$=3700 K Ω_2=7.607   F$_1$=1.00

T$_2$=3664   q=0.985   F$_2$=1.00

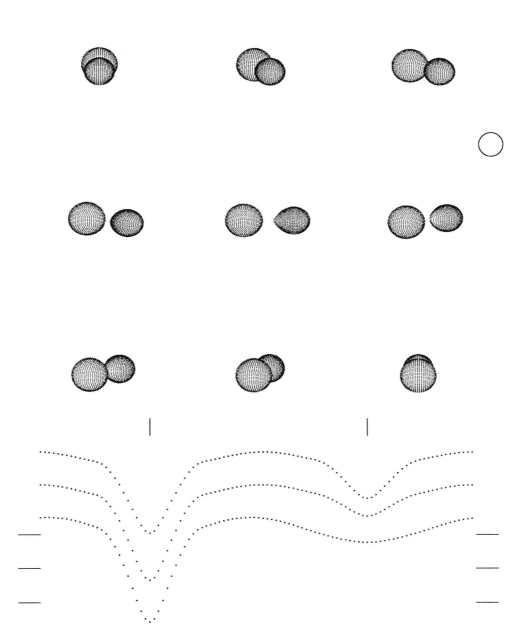

a=4.1 R$_\odot$ r$_1$(pole)=0.350   r$_2$(pole)=0.289

e= 0.000   r$_1$(point)=0.380   r$_2$(point)=0.416

ω= ---   r$_1$(side)=0.361   r$_2$(side)=0.302

P=0^d.6594   r$_1$(back)=0.372   r$_2$(back)=0.334

i=81°.41 Ω_1=3.271   V$_\gamma$= ---

T$_1$=7700 K Ω_2=2.751   F$_1$=1.00

T$_2$=5270 K   q=0.436   F$_2$=1.00

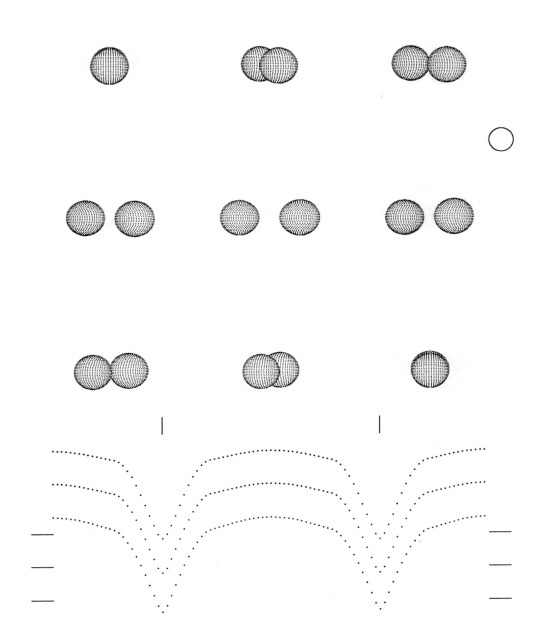

a=5.0 R$_\odot$ r$_1$(pole)=0.298   r$_2$(pole)=0.305

e= 0.000   r$_1$(point)=0.328   r$_2$(point)=0.339

ω= ---   r$_1$(side)=0.306   r$_2$(side)=0.314

P=0$^\mathrm{d}$.7161   r$_1$(back)=0.319   r$_2$(back)=0.328

i=88°.3 Ω_1=4.318   V$_\gamma$= ---

T$_1$=8000 K Ω_2=4.237   F$_1$=1.00

T$_2$=7990 K   q=1.0   F$_2$=1.00

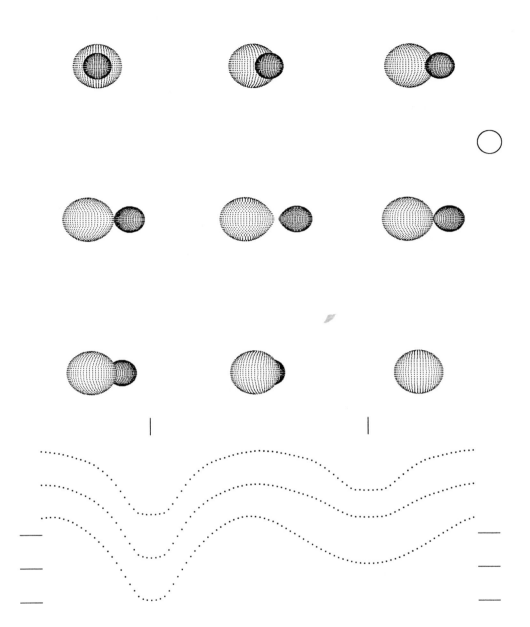

a= 4.23 $R_\odot$ r_1(pole)=0.444 r_2(pole)=0.263

e= 0.000 r_1(point)=0.558 r_2(point)=0.348

ω= --- r_1(side)=0.475 r_2(side)=0.273

P=0^d.6240 r_1(back)=0.498 r_2(back)=0.304

i=$90°$.0 Ω_1=2.546 V_γ=+3.6 km sec^{-1}

T_1=7200 K Ω_2=2.530 F_1=1.0

T_2=4659 K   q=0.32 F_2=1.0

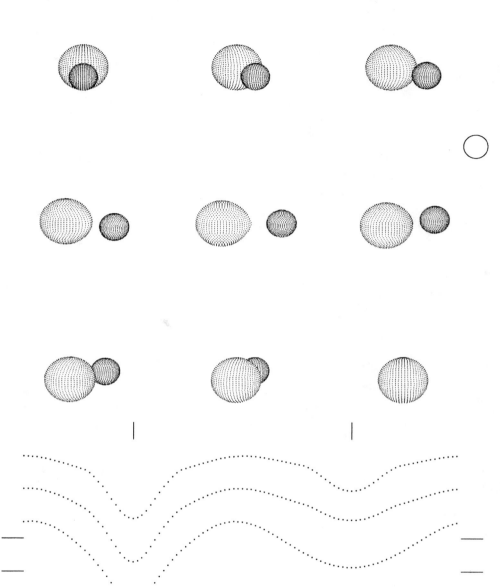

a= 4.9 R$_\odot$	r$_1$(pole)=0.396	r$_2$(pole)=0.237
e= 0.000	r$_1$(point)=0.491	r$_2$(point)=0.256
ω= ---	r$_1$(side)=0.418	r$_2$(side)=0.242
P=0^d.7121	r$_1$(back)=0.443	r$_2$(back)=0.251
i=80°.3	Ω_1=3.034	V$_\gamma$= ---
T$_1$=8100 K	Ω_2=3.515	F$_1$=1.00
T$_2$=4750 K	q=0.55	F$_2$=1.00

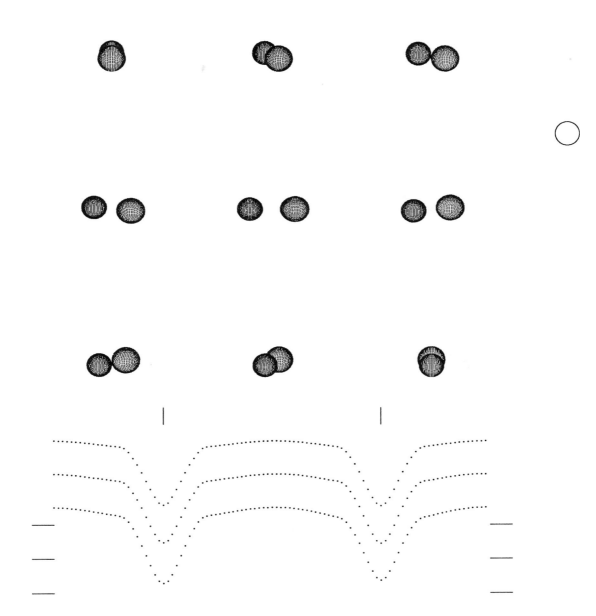

a= 3.7 R$_\odot$ r$_1$(pole)=0.268   r$_2$(pole)=0.296

e= .000   r$_1$(point)=0.284   r$_2$(point)=0.329

ω= ---   r$_1$(side)=0.273   r$_2$(side)=0.305

P=0^d.6001   r$_1$(back)=0.280   r$_2$(back)=0.318

i=84°.00 Ω_1=4.632   V$_\gamma$=−18.0 km sec^{-1}

T$_1$=6000 K Ω_2=4.130   F$_1$=1.00

T$_2$=5930 K   q=0.930   F$_2$=1.00

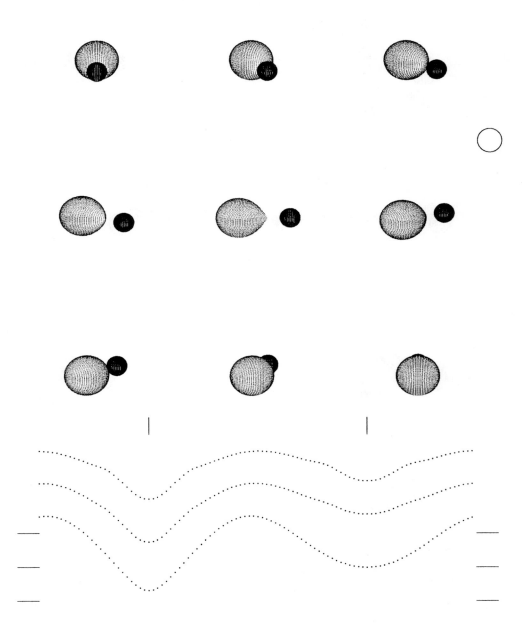

a= 4.20 $R_\odot$ r_1(pole)=0.404 r_2(pole)=0.196

e= 0.000 r_1(point)=0.533 r_2(point)=0.203

ω= − − − r_1(side)=0.428 r_2(side)=0.198

P=$0^d.8830$ r_1(back)=0.457 r_2(back)=0.202

i=$76°.34$ Ω_1=2.991 V_γ= +5 km sec^{-1}

T_1=7965 K Ω_2=4.060 F_1=1.00

T_2=4690 K   q=0.56 F_2=1.00

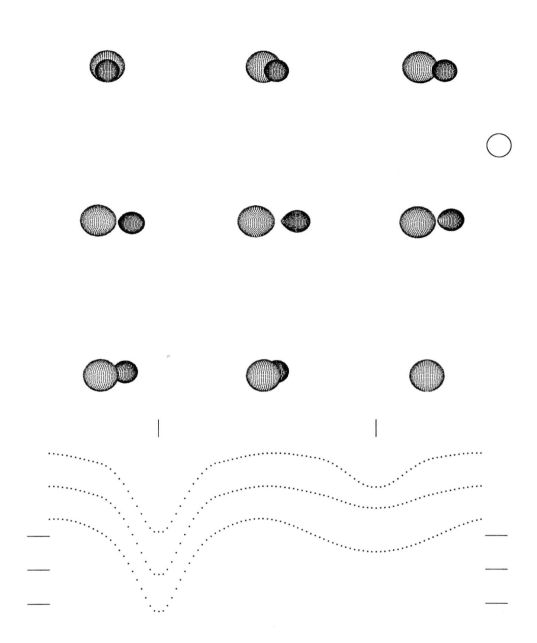

$$a=3.40\ R_\odot \qquad r_1(\text{pole})=0.404 \qquad r_2(\text{pole})=0.279$$

$$e=\ 0.000 \qquad r_1(\text{point})=0.460 \qquad r_2(\text{point})=0.402$$

$$\omega=\ \text{---} \qquad r_1(\text{side})=0.421 \qquad r_2(\text{side})=0.290$$

$$P=0^d.6441 \qquad r_1(\text{back})=0.437 \qquad r_2(\text{back})=0.323$$

$$i=84^\circ.0 \qquad \Omega_1=2.850 \qquad V_\gamma=23.8\ \text{km sec}^{-1}$$

$$T_1=7230\ K \qquad \Omega_2=2.637 \qquad F_1=1.00$$

$$T_2=3800\ K \qquad q=0.380 \qquad F_2=1.00$$

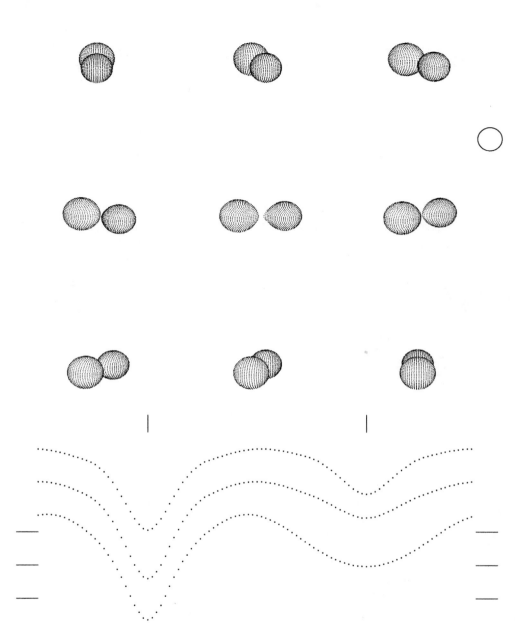

a= 3.8 R$_\odot$	r$_1$(pole)=0.367	r$_2$(pole)=0.327
e= 0.000	r$_1$(point)=0.439	r$_2$(point)=0.463
ω= ---	r$_1$(side)=0.384	r$_2$(side)=0.342
P=0^d.5288	r$_1$(back)=0.407	r$_2$(back)=0.374
i=78°.3	Ω_1=3.381	V$_\gamma$= ---
T$_1$=6800 K	Ω_2=3.243	F$_1$=1.00
T$_2$=4160 K	q=0.70	F$_2$=1.00

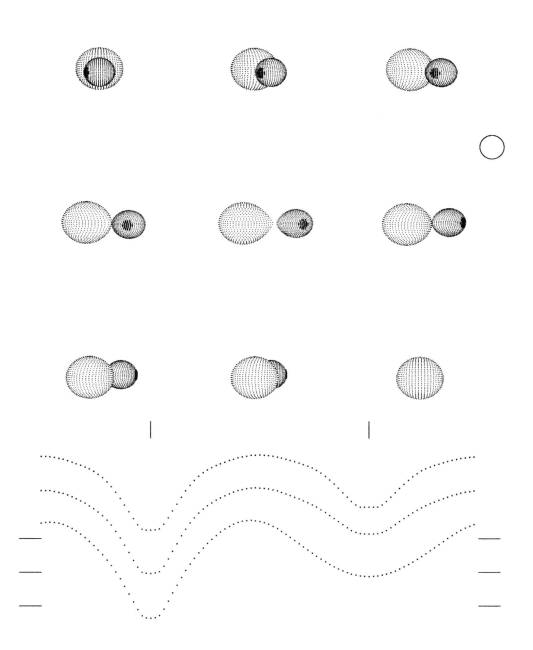

a= 4.29 R$_\odot$ r$_1$(pole)=0.424   r$_2$(pole)=0.286

e= 0.000   r$_1$(point)=0.582   r$_2$(point)=0.373

ω= ———   r$_1$(side)=0.451   r$_2$(side)=0.298

P=0^d.6614   r$_1$(back)=0.478   r$_2$(back)=0.328

i=86°.05 Ω_1=2.771   V$_\gamma$=−41 km sec^{-1}

T$_1$=7500 K Ω_2=2.795   F$_1$=1.0

T$_2$=5200 K   q=0.446   F$_2$=1.0

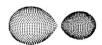

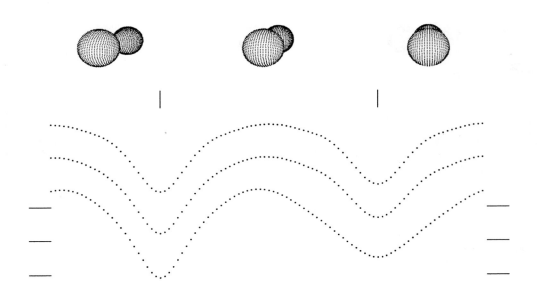

$a=$ 4.0 $R_\odot$ r_1(pole)=0.406 r_2(pole)=0.298

$e=$ 0.000 r_1(point)=0.545 r_2(point)=0.374

$\omega=$ --- r_1(side)=0.430 r_2(side)=0.310

$P=0^d.5804$ r_1(back)=0.458 r_2(back)=0.338

$i=79°.1$ $\Omega_1=2.974$ $V_\gamma=$ ---

$T_1=7000$ K $\Omega_2=3.027$ $F_1=1.00$

$T_2=5900$ K $q=0.55$ $F_2=1.00$

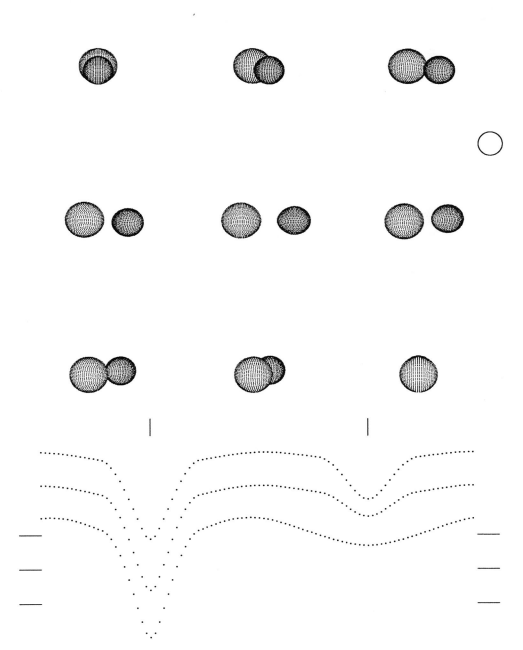

a= 4.6 $R_\odot$	r_1(pole)=0.344	r_2(pole)=0.273
e= 0.000	r_1(point)=0.374	r_2(point)=0.322
ω= ---	r_1(side)=0.355	r_2(side)=0.282
P=0^d.7118	r_1(back)=0.365	r_2(back)=0.303
i=$84°$.87	Ω_1=3.358	V_γ= ---
T_1=7800 K	Ω_2=2.961	F_1=1.00
T_2=5098 K	q=0.473	F_2=1.00

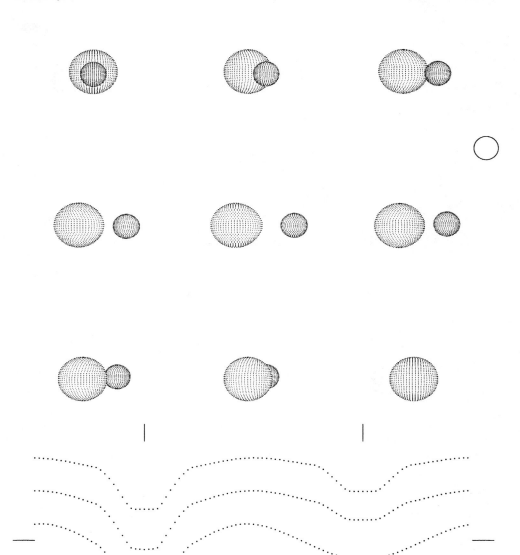

a=4.8 R$_\odot$ r$_1$(pole)=0.394   r$_2$(pole)=0.218

e= 0.000   r$_1$(point)=0.458   r$_2$(point)=0.234

ω= −−−   r$_1$(side)=0.414   r$_2$(side)=0.222

P=0$^\mathrm{d}$.7553   r$_1$(back)=0.432   r$_2$(back)=0.230

i=87°.5 Ω_1=2.944   V$_\gamma$= −−−

T$_1$=7560 K Ω_2=3.265   F$_1$=1.00

T$_2$=5240 K   q=0.438   F$_2$=1.00

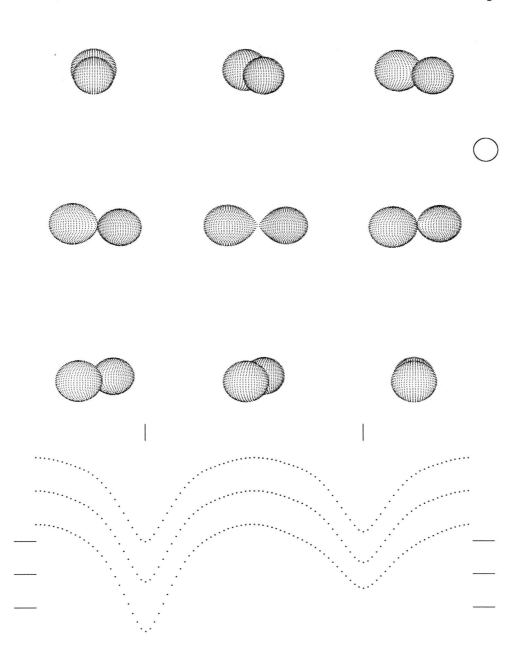

$a=4.74\ R_{\odot}$ $r_1(pole)=0.374$ $r_2(pole)=0.337$

$e=\ 0.000$ $r_1(point)=0.505$ $r_2(point)=0.459$

$\omega=\ ---$ $r_1(side)=0.395$ $r_2(side)=0.354$

$P=0^d.7949$ $r_1(back)=0.424$ $r_2(back)=0.384$

$i=84°.0$ $\Omega_1=3.416$ $V_{\gamma}=+25.0\ km\ sec^{-1}$

$T_1=8150\ K$ $\Omega_2=3.416$ $F_1=1.00$

$T_2=7390\ K$ $q=0.8$ $F_2=1.00$

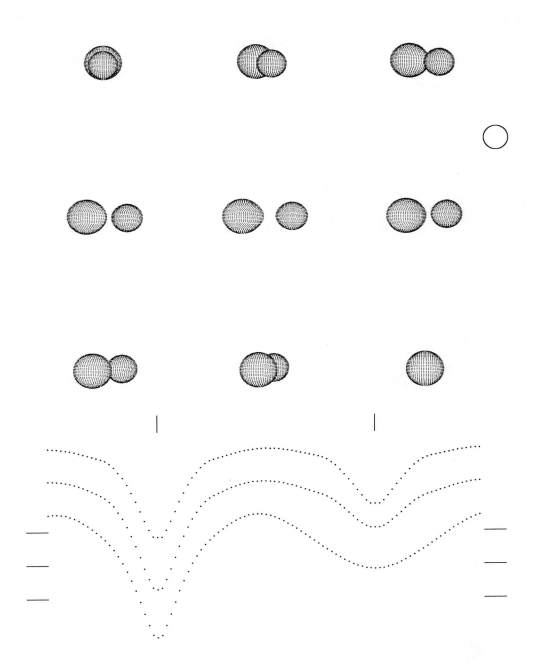

a= 4.1 $R_\odot$ r_1(pole)=0.360 r_2(pole)=0.291

e= 0.000 r_1(point)=0.435 r_2(point)=0.327

ω= ——— r_1(side)=0.377 r_2(side)=0.300

P=0^d.6642 r_1(back)=0.401 r_2(back)=0.315

i=87°.32 Ω_1=3.501 V_γ= ———

T_1=7000 K Ω_2=3.734 F_1=1.00

T_2=4749 K   q=0.774 F_2=1.00

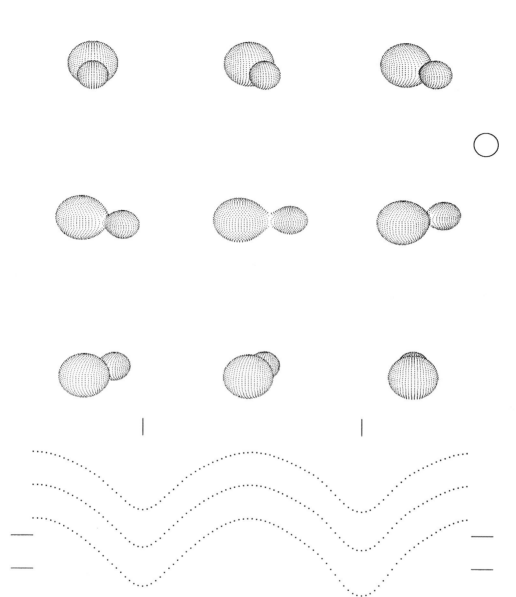

a=4.09 R$_\odot$ r$_1$(pole)=0.462   r$_2$(pole)=0.294

e= 0.000   r$_1$(point)=−1.000   r$_2$(point)=−1.000

ω= — — —   r$_1$(side)=0.500   r$_2$(side)=0.310

P=0^d.5948   r$_1$(back)=0.534   r$_2$(back)=0.362

i=76°.1 Ω_1=2.486   V$_\gamma$=20.6 km sec^{-1}

T$_1$=6900 K Ω_2=2.486   F$_1$=1.00

T$_2$=7030 K   q=0.354   F$_2$=1.00

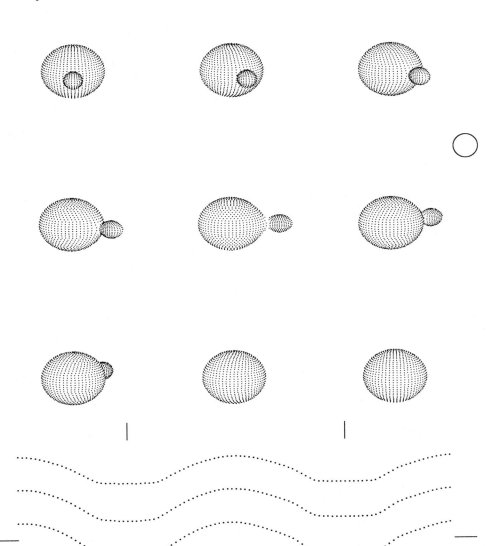

a=4.01 R$_\odot$ r$_1$(pole)=0.567   r$_2$(pole)=0.189

e= 0.000   r$_1$(point)=−1.000   r$_2$(point)=−1.000

ω= ---   r$_1$(side)=0.649   r$_2$(side)=0.199

P=0^d.4387   r$_1$(back)=0.668   r$_2$(back)=0.269

i=79°.1 Ω_1=1.826   V$_\gamma$=−17.0 km sec^{-1}

T$_1$=7175 K Ω_2=1.826   F$_1$=1.00

T$_2$=6910 K   q=0.072   F$_2$=1.00

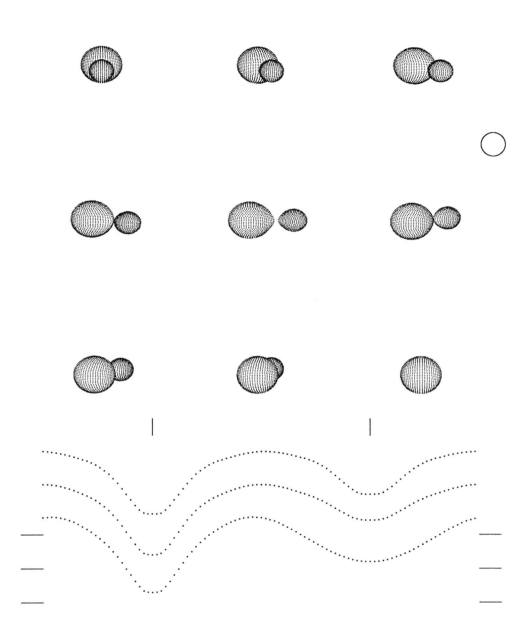

$$a=3.6\ R_\odot \qquad r_1(\text{pole})=0.439 \qquad r_2(\text{pole})=0.269$$

$$e=0.000 \qquad r_1(\text{point})=0.559 \qquad r_2(\text{point})=0.359$$

$$\omega=--- \qquad r_1(\text{side})=0.469 \qquad r_2(\text{side})=0.280$$

$$P=0^d.5249 \qquad r_1(\text{back})=0.494 \qquad r_2(\text{back})=0.269$$

$$i=81°.6 \qquad \Omega_1=2.594 \qquad V_\gamma=---$$

$$T_1=7100\ K \qquad \Omega_2=2.583 \qquad F_1=1.00$$

$$T_2=4840\ K \qquad q=0.348 \qquad F_2=1.00$$

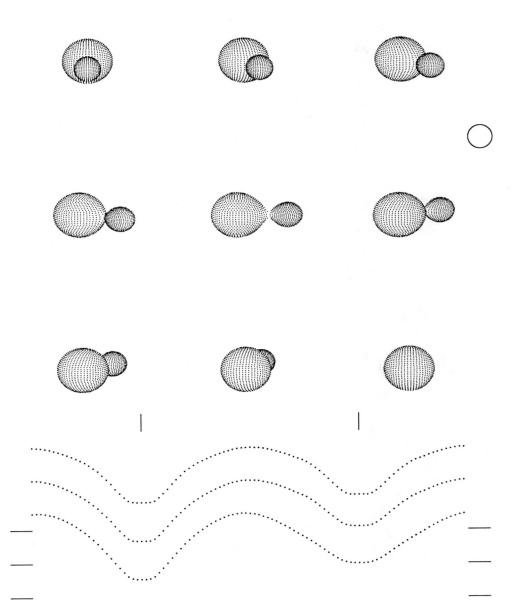

$$a=4.11 \; R_\odot \qquad r_1(\text{pole})=0.466 \qquad r_2(\text{pole})=0.260$$

$$e= \; 0.000 \qquad r_1(\text{point})=-1.000 \qquad r_2(\text{point})=-1.000$$

$$\omega= \; \text{---} \qquad r_1(\text{side})=0.504 \qquad r_2(\text{side})=0.271$$

$$P=0^d.6427 \qquad r_1(\text{back})=0.530 \qquad r_2(\text{back})=0.308$$

$$i=80°.7 \qquad \Omega_1=2.395 \qquad V_\gamma=-5.6 \; \text{km sec}^{-1}$$

$$T_1=7700 \; K \qquad \Omega_2=2.395 \qquad F_1=1.00$$

$$T_2=6520 \; K \qquad q=0.277 \qquad F_2=1.00$$

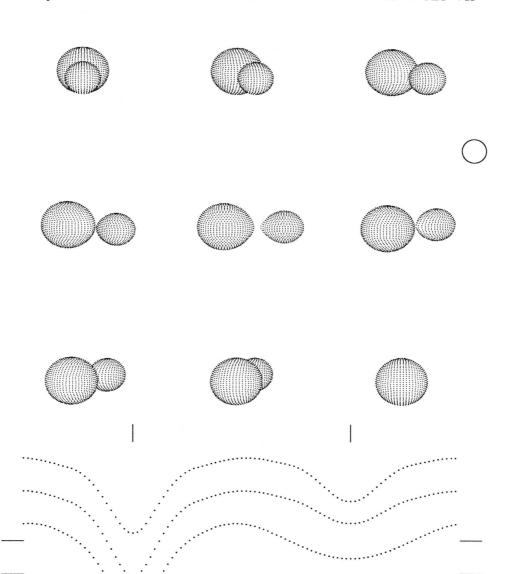

a=4.9 R$_\odot$ r$_1$(pole)=0.412   r$_2$(pole)=0.286

e= 0.000   r$_1$(point)=0.496   r$_2$(point)=0.412

ω= ---   r$_1$(side)=0.436   r$_2$(side)=0.298

P=0^d.7025   r$_1$(back)=0.457   r$_2$(back)=0.331

i=82°.6 Ω_1=2.815   V$_\gamma$= ---

T$_1$=9600 K Ω_2=2.719   F$_1$=1.00

T$_2$=5624 K   q=0.420   F$_2$=1.00

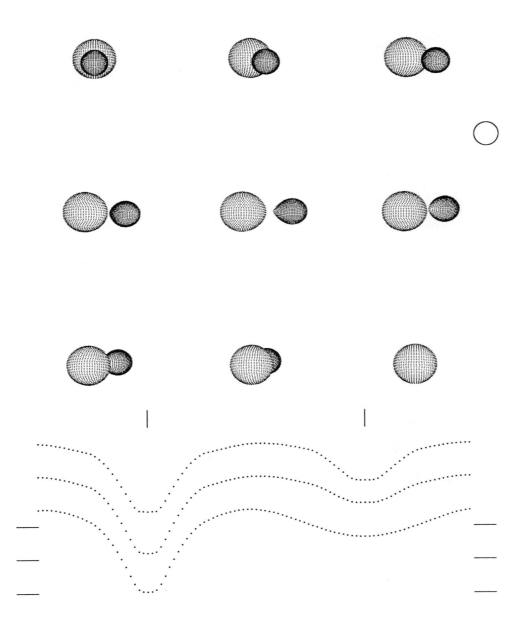

a=4.07 $R_\odot$ r_1(pole)=0.414 r_2(pole)=0.268

e= 0.000 r_1(point)=0.477 r_2(point)=0.389

ω= ––– r_1(side)=0.436 r_2(side)=0.279

P=0^d.6406 r_1(back)=0.453 r_2(back)=0.312

i=$85°$.51 Ω_1=2.719 V_γ=9.0 km sec^{-1}

T_1=9750 K Ω_2=2.534 F_1=1.00

T_2=5560 K   q=0.331 F_2=1.00

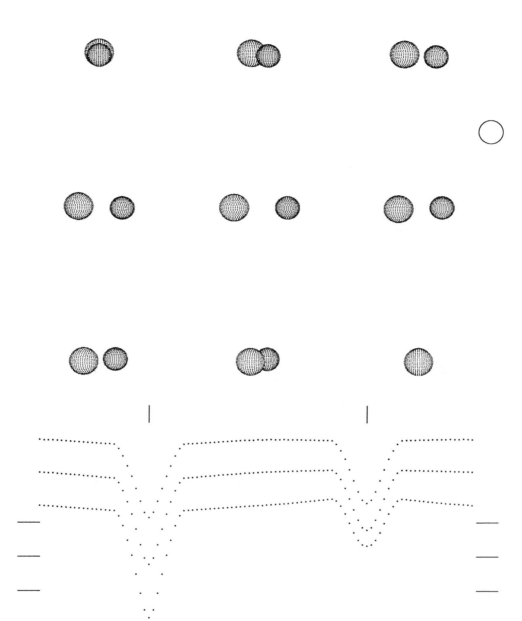

a=4.41 $R_\odot$ r_1(pole)=0.263 r_2(pole)=0.223

e= 0.000 r_1(point)=0.278 r_2(point)=0.232

ω= – – – r_1(side)=0.268 r_2(side)=0.226

P=0^d.8169 r_1(back)=0.275 r_2(back)=0.230

i=87°.08 Ω_1=4.622 V_γ=−28.7 km sec^{-1}

T_1=6250 K Ω_2=4.900 F_1=1.00

T_2=5625 K   q=0.859 F_2=1.00

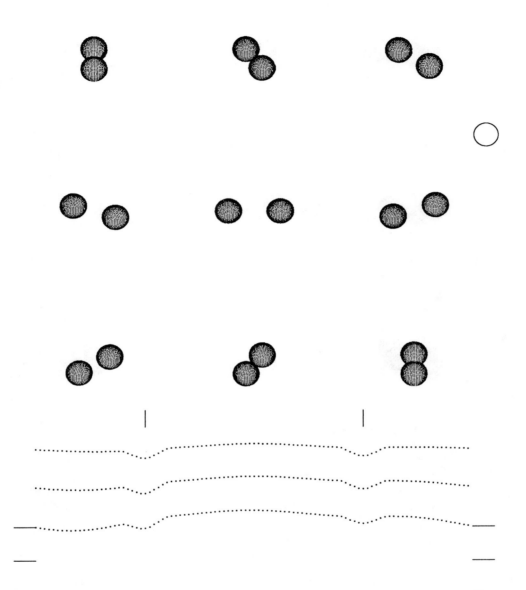

a= 4.28 $R_\odot$ r_1(pole)=0.250 r_2(pole)=0.250

e= 0.000 r_1(point)=0.262 r_2(point)=0.263

ω= ——— r_1(side)=0.254 r_2(side)=0.254

P=0^d.6981 r_1(back)=0.260 r_2(back)=0.260

i=66°.7 Ω_1=4.928 V_γ=−24.6 km sec^{-1}

T_1=5900 K Ω_2=4.818 F_1=1.0

T_2=5750 K   q=0.957 F_2=1.0

Group IV
Systems with
$5.06\ R_\odot < a \leqslant 7.60\ R_\odot$

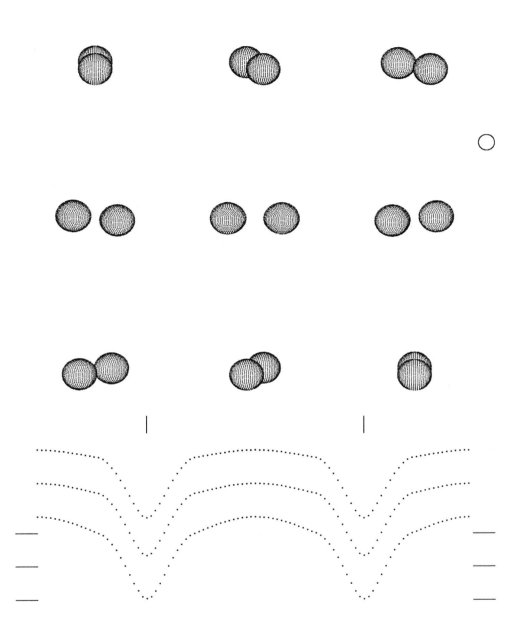

a= 6.7 $R_\odot$ r_1(pole)=0.307 r_2(pole)=0.305

e= 0.000 r_1(point)=0.342 r_2(point)=0.339

ω= ——— r_1(side)=0.316 r_2(side)=0.314

P=0^d.9862 r_1(back)=0.330 r_2(back)=0.328

i=81°.9 Ω_1=4.218 V_γ= ———

T_1=9500 K Ω_2=4.237 F_1=1.00

T_2=9500 K   q=1.0 F_2=1.00

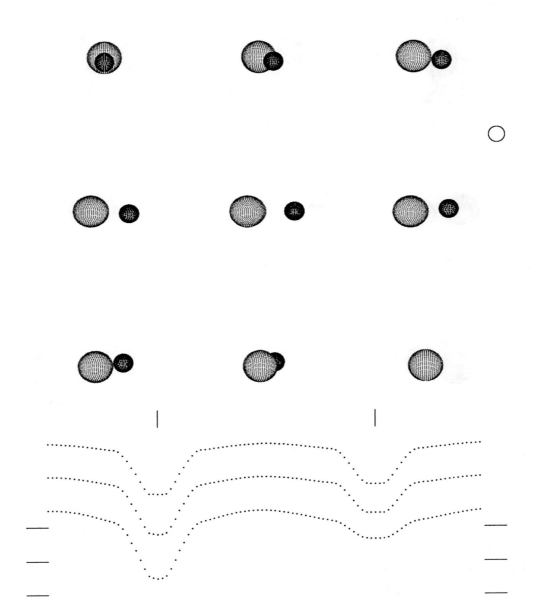

$$a=5.82\ R_\odot \qquad r_1(pole)=0.352 \qquad r_2(pole)=0.204$$

$$e=0.000 \qquad r_1(point)=0.395 \qquad r_2(point)=0.212$$

$$\omega=\ \text{---} \qquad r_1(side)=0.365 \qquad r_2(side)=0.207$$

$$P=1^d.0105 \qquad r_1(back)=0.380 \qquad r_2(back)=0.211$$

$$i=84^\circ.3 \qquad \Omega_1=3.400 \qquad V_\gamma=5.4\ km\ sec^{-1}$$

$$T_1=6775\ K \qquad \Omega_2=4.090 \qquad F_1=1.00$$

$$T_2=5997\ K \qquad q=0.593 \qquad F_2=1.00$$

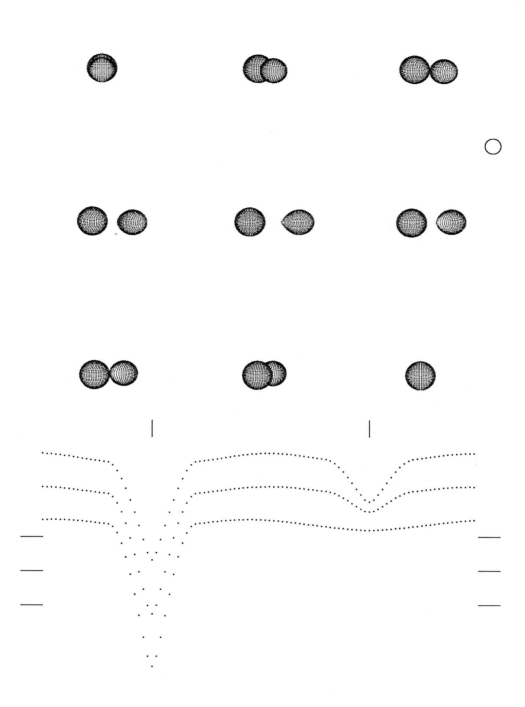

a= 6.1 R$_\odot$ r$_1$(pole)=0.298   r$_2$(pole)=0.261

e= 0.000   r$_1$(point)=0.308   r$_2$(point)=0.379

ω= ———   r$_1$(side)=0.303   r$_2$(side)=0.272

P=1^d.1064   r$_1$(back)=0.307   r$_2$(back)=0.305

i=87°.7 Ω_1=3.644   V$_\gamma$= ———

T$_1$=9500 K Ω_2=2.466   F$_1$=1.00

T$_2$=5420 K   q=0.3   F$_2$=1.00

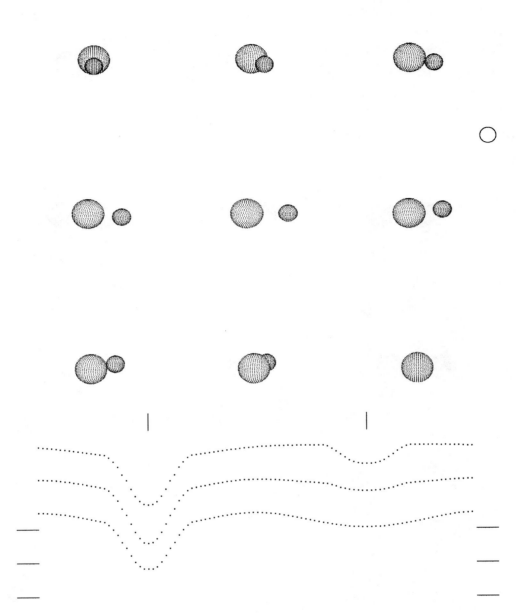

a=5.1 R$_\odot$ r$_1$(pole)=0.371   r$_2$(pole)=0.211

e= 0.000   r$_1$(point)=0.400   r$_2$(point)=0.235

ω= ———   r$_1$(side)=0.385   r$_2$(side)=0.216

P=1^d.0296   r$_1$(back)=0.393   r$_2$(back)=0.228

i=81°.33 Ω_1=2.950   V$_\gamma$= ———

T$_1$=6800 K Ω_2=2.641   F$_1$=1.00

T$_2$=3030 K   q=0.2748   F$_2$=1.00

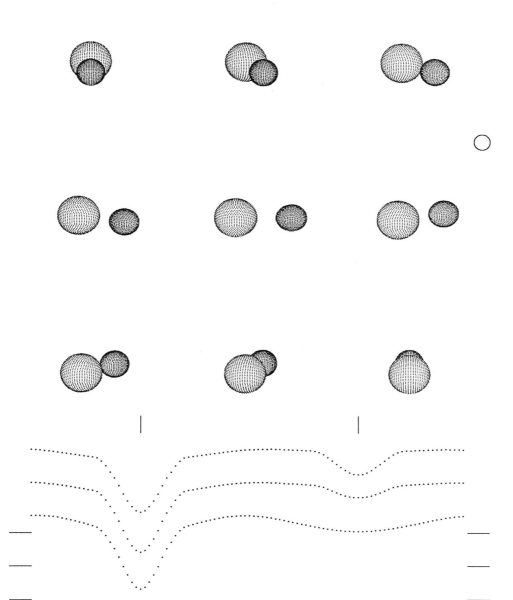

a=6.95 $R_{\odot}$ r_1(pole)=0.357 r_2(pole)=0.246

e= 0.000 r_1(point)=0.385 r_2(point)=0.286

ω= – – – r_1(side)=0.369 r_2(side)=0.253

P=1^d.2473 r_1(back)=0.378 r_2(back)=0.272

i=78°.3 Ω_1=3.125 V_{γ}=+16.8 km sec^{-1}

T_1=10350 K Ω_2=2.704 F_1=1.00

T_2=5031 K   q=0.345 F_2=1.00

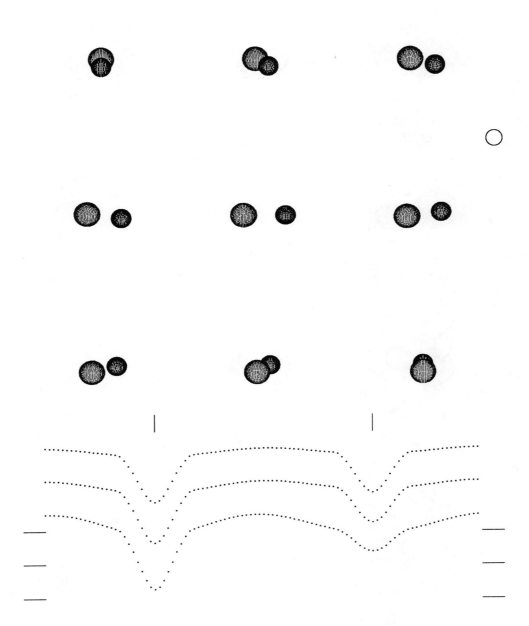

a=5.2 R$_\odot$ r$_1$(pole)=0.297   r$_2$(pole)=0.232

e= 0.000   r$_1$(point)=0.320   r$_2$(point)=0.243

ω= ———   r$_1$(side)=0.304   r$_2$(side)=0.235

P=0^d.8425   r$_1$(back)=0.313   r$_2$(back)=0.240

i=79°.27 Ω_1=4.132   V$_\gamma$= ———

T$_1$=7188 K Ω_2=4.505   F$_1$=1.00

T$_2$=6195 K   q=0.794   F$_2$=1.00

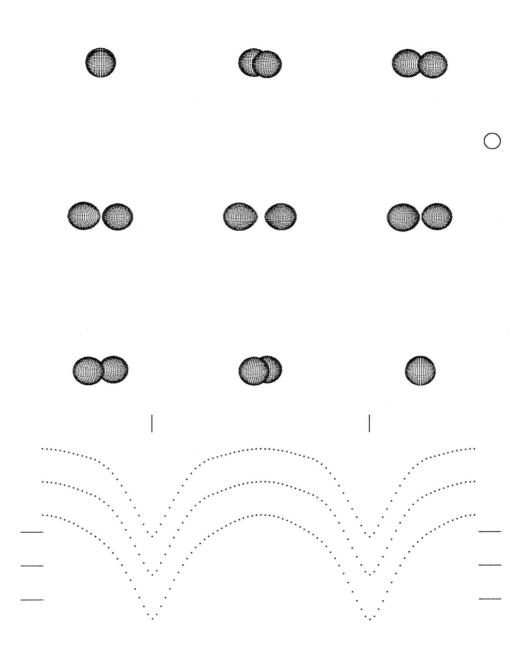

a=5.15 R$_\odot$ r$_1$(pole)=0.348   r$_2$(pole)=0.334

e= 0.000   r$_1$(point)=0.436   r$_2$(point)=0.395

ω= − − −   r$_1$(side)=0.364   r$_2$(side)=0.347

P=1^d.0826   r$_1$(back)=0.391   r$_2$(back)=0.369

i=87°.6 Ω_1=3.820   V$_\gamma$=15.0 km sec^{-1}

T$_1$=9500 K Ω_2=3.944   F$_1$=1.00

T$_2$=9500 K   q=1.000   F$_2$=1.00

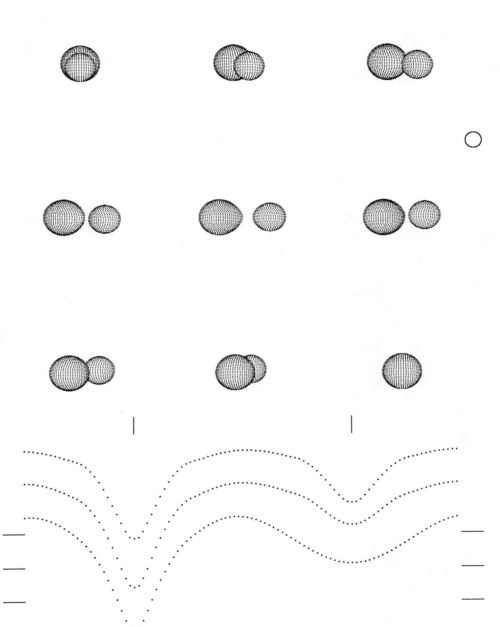

$a=6.2\ R_{\odot}$ $r_1(pole)=0.371$ $r_2(pole)=0.295$

$e=\ 0.000$ $r_1(point)=0.461$ $r_2(point)=0.335$

$\omega=\ ---$ $r_1(side)=0.389$ $r_2(side)=0.304$

$P=0^d.9016$ $r_1(back)=0.415$ $r_2(back)=0.321$

$i=86^{\circ}.0$ $\Omega_1=3.402$ $V_{\gamma}=\ ---$

$T_1=8800\ K$ $\Omega_2=3.630$ $F_1=1.00$

$T_2=5580\ K$ $q=0.750$ $F_2=1.00$

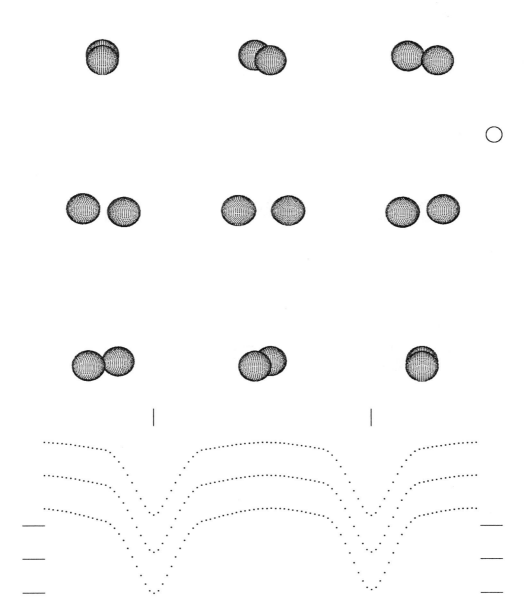

a=6.2 R$_\odot$	r$_1$(pole)=0.316	r$_2$(pole)=0.308
e= 0.000	r$_1$(point)=0.352	r$_2$(point)=0.348
ω= ---	r$_1$(side)=0.325	r$_2$(side)=0.317
P=1^d.2234	r$_1$(back)=0.340	r$_2$(back)=0.334
i=83°.7	Ω_1=4.030	V$_\gamma$= ---
T$_1$=6210 K	Ω_2=3.940	F$_1$=1.00
T$_2$=6167 K	q=0.900	F$_2$=1.00

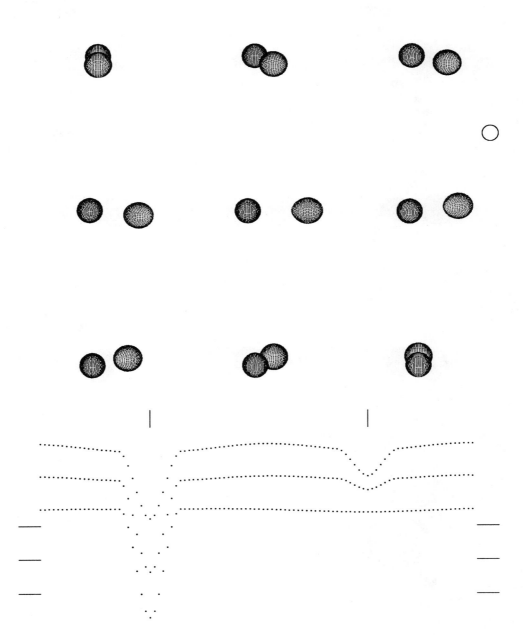

a= 7.41 R$_\odot$ r$_1$(pole)=0.212   r$_2$(pole)=0.223

e= 0.000   r$_1$(point)=0.214   r$_2$(point)=0.270

ω= ---   r$_1$(side)=0.213   r$_2$(side)=0.230

P=2^d.3373   r$_1$(back)=0.214   r$_2$(back)=0.253

i=81°.31 Ω_1=4.930   V$_\gamma$= −29 km sec^{-1}

T$_1$=7616 K Ω_2=2.322   F$_1$=1.00

T$_2$=4464 K   q=0.212   F$_2$=1.00

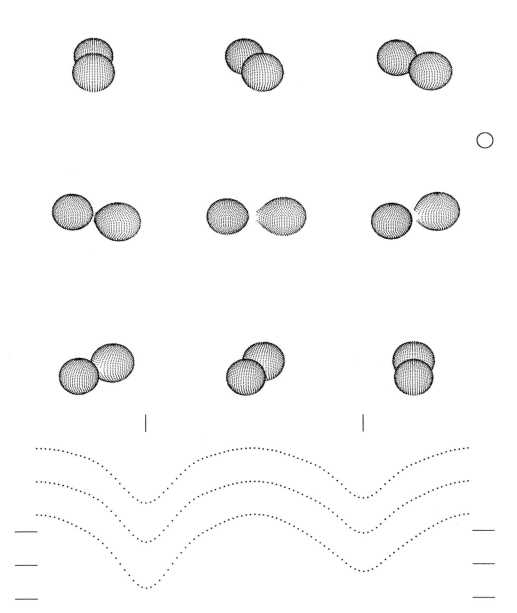

a= 6.96 $R_\odot$	r_1(pole)=0.334	r_2(pole)=0.355
e= 0.000	r_1(point)=0.385	r_2(point)=−1.000
ω= −−−	r_1(side)=0.346	r_2(side)=0.374
P=0^d.9362	r_1(back)=0.365	r_2(back)=0.410
i=72°.45	Ω_1=3.840	V_γ=−22 km sec^{-1}
T_1=10000 K	Ω_2=3.504	F_1=1.0
T_2=9500 K	q=0.89	F_2=1.0

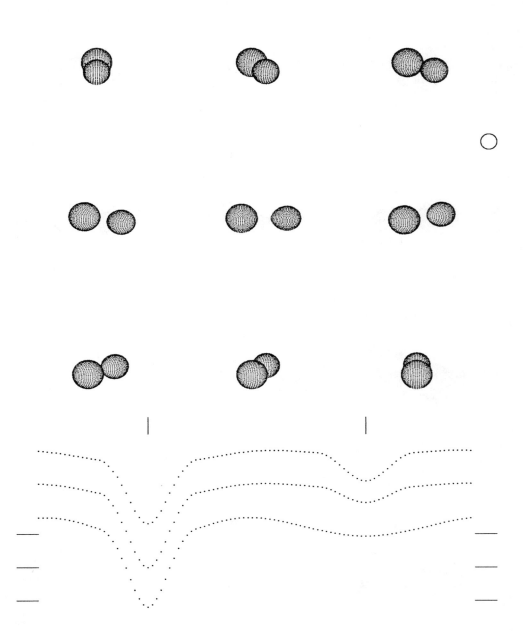

a= 5.6 R$_\odot$	r$_1$(pole)=0.333	r$_2$(pole)=0.286
e= 0.000	r$_1$(point)=0.360	r$_2$(point)=0.349
ω= ---	r$_1$(side)=0.343	r$_2$(side)=0.296
P=0^d.6867	r$_1$(back)=0.352	r$_2$(back)=0.321
i=77°.28	Ω_1=3.475	V$_\gamma$= ---
T$_1$=11000 K	Ω_2=2.961	F$_1$=1.00
T$_2$=5371 K	q=0.50	F$_2$=1.00

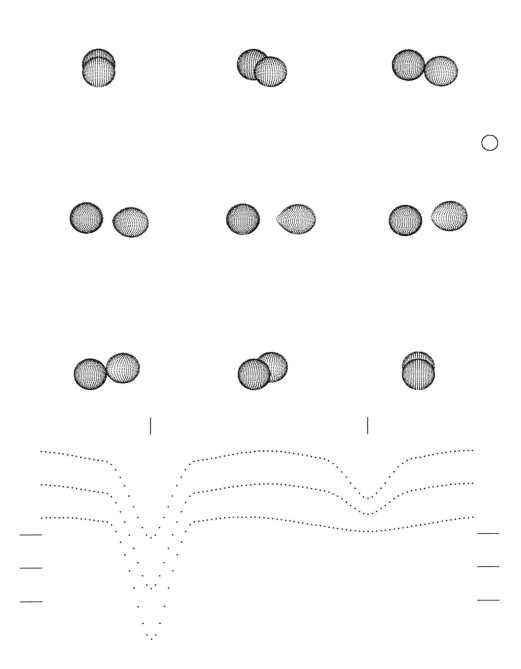

a= 6.8 $R_\odot$ r_1(pole)=0.292 r_2(pole)=0.282

e= 0.000 r_1(point)=0.304 r_2(point)=0.407

ω= --- r_1(side)=0.297 r_2(side)=0.294

P=1^d.3326 r_1(back)=0.301 r_2(back)=0.327

i=81°.9 Ω_1=3.811 V_γ= ---

T_1=8800 K Ω_2=2.678 F_1=1.00

T_2=5530 K   q=0.4 F_2=1.00

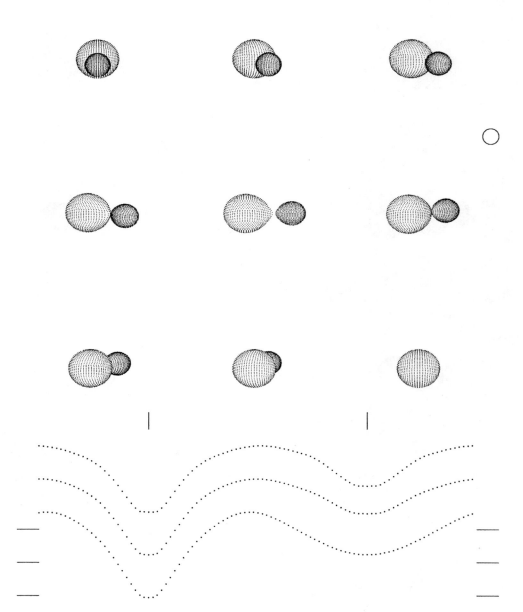

a= 5.6 $R_\odot$ r_1(pole)=0.442 r_2(pole)=0.270

e= 0.000 r_1(point)=0.578 r_2(point)=0.351

ω= --- r_1(side)=0.472 r_2(side)=0.281

P=0^d.8590 r_1(back)=0.358 r_2(back)=0.311

i=82°.4 Ω_1=3.591 V_γ= ---

T_1=9600 K Ω_2=2.610 F_1=1.00

T_2=5690 K   q=0.358 F_2=1.00

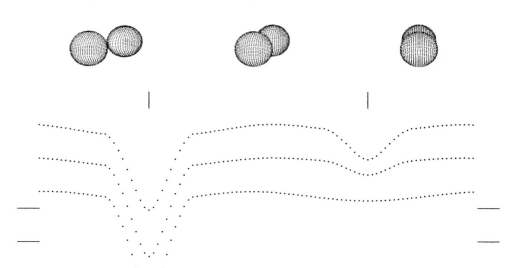

a= 7.45 $R_\odot$ r_1(pole)=0.305 r_2(pole)=0.270

e= 0.000 r_1(point)=0.318 r_2(point)=0.391

ω= ——— r_1(side)=0.311 r_2(side)=0.282

P=1^d.1988 r_1(back)=0.315 r_2(back)=0.314

i=81°.1 Ω_1=3.602 V_γ=−1.3 km sec^{-1}

T_1=10520 K Ω_2=2.553 F_1=1.00

T_2=5180 K   q=0.43 F_2=1.00

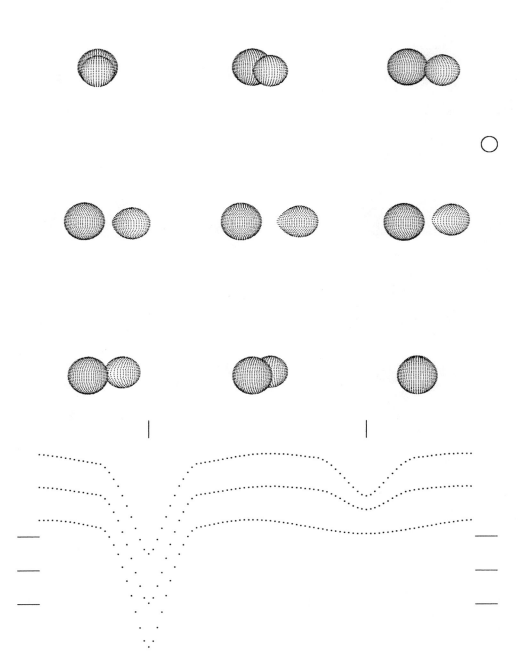

$a = 7.2\ R_\odot$ $r_1(\text{pole}) = 0.332$ $r_2(\text{pole}) = 0.281$

$e = 0.000$ $r_1(\text{point}) = 0.353$ $r_2(\text{point}) = 0.405$

$\omega = \ ---$ $r_1(\text{side}) = 0.341$ $r_2(\text{side}) = 0.292$

$P = 1^d.2435$ $r_1(\text{back}) = 0.348$ $r_2(\text{back}) = 0.325$

$i = 86°.0$ $\Omega_1 = 3.383$ $V_\gamma = \ ---$

$T_1 = 11000\ K$ $\Omega_2 = 2.658$ $F_1 = 1.00$

$T_2 = 5600\ K$ $q = 0.390$ $F_2 = 1.00$

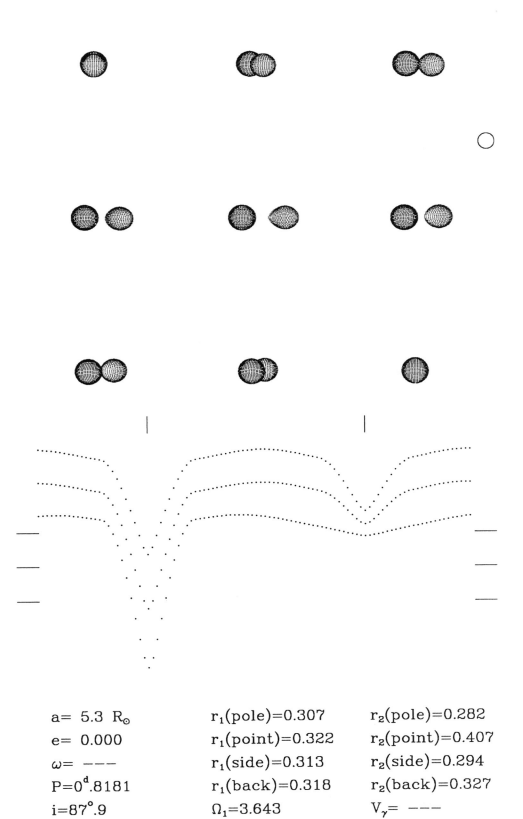

a= 5.3 R$_\odot$

e= 0.000

ω= ———

P=0^d.8181

i=87°.9

T$_1$=8200 K

T$_2$=5600 K

r$_1$(pole)=0.307

r$_1$(point)=0.322

r$_1$(side)=0.313

r$_1$(back)=0.318

Ω_1=3.643

Ω_2=2.678

q=0.4

r$_2$(pole)=0.282

r$_2$(point)=0.407

r$_2$(side)=0.294

r$_2$(back)=0.327

V$_\gamma$= ———

F$_1$=1.00

F$_2$=1.00

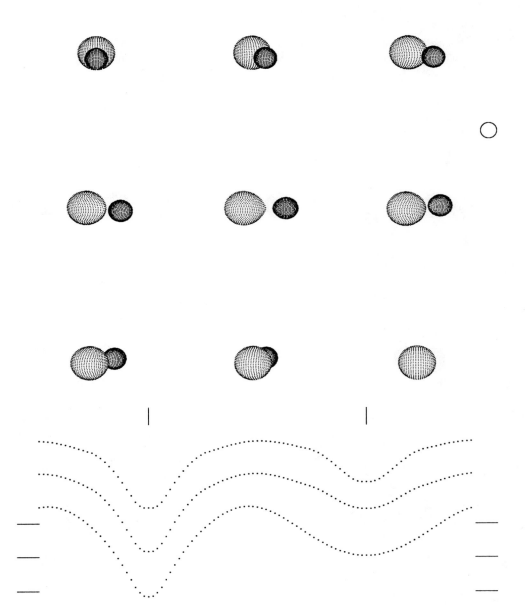

a= 5.18 R$_\odot$ r$_1$(pole)=0.417   r$_2$(pole)=0.266

e= 0.000   r$_1$(point)=0.522   r$_2$(point)=0.313

ω= ---   r$_1$(side)=0.443   r$_2$(side)=0.275

P=0^d.9121   r$_1$(back)=0.467   r$_2$(back)=0.296

i=82°.31 Ω_1=2.801   V$_\gamma$= +21.8 km sec^{-1}

T$_1$=7239 K Ω_2=2.896   F$_1$=1.00

T$_2$=4690 K   q=0.439   F$_2$=1.00

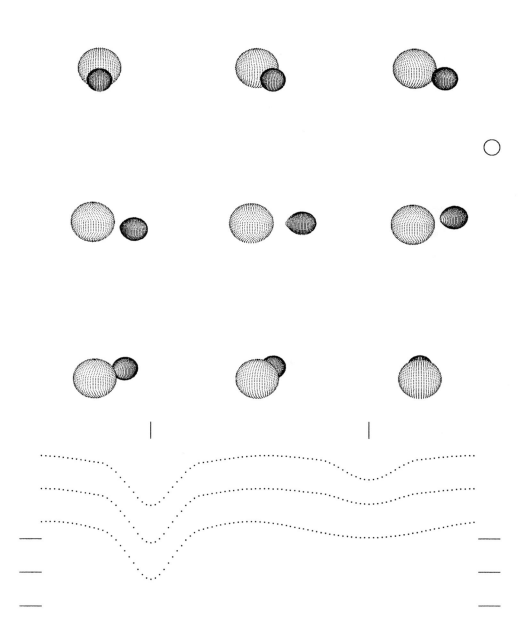

a= 6.3 $R_{\odot}$ r_1(pole)=0.403 r_2(pole)=0.233

e= 0.000 r_1(point)=0.438 r_2(point)=0.341

ω= ——— r_1(side)=0.421 r_2(side)=0.242

P=1^d.0840 r_1(back)=0.429 r_2(back)=0.275

i=75°.56 Ω_1=2.665 V_{γ}= ———

T_1=9400 K Ω_2=2.233 F_1=1.00

T_2=5063 K   q=0.20 F_2=1.00

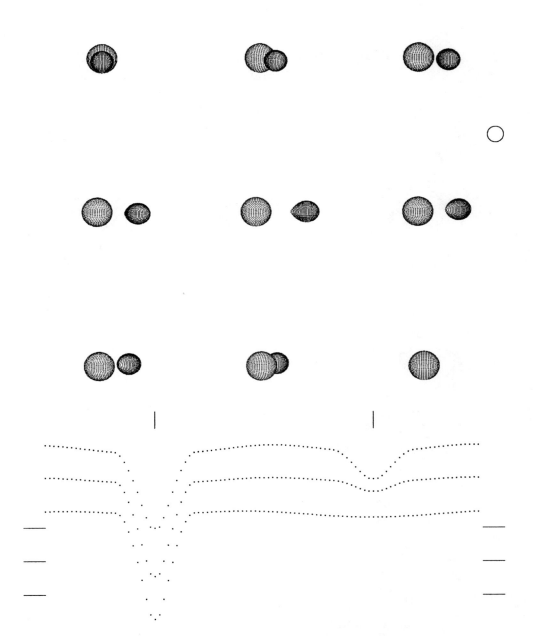

$a= 6.2\ R_\odot$ $r_1(pole)=0.299$ $r_2(pole)=0.218$

$e= 0.000$ $r_1(point)=0.307$ $r_2(point)=0.322$

$\omega= \text{---}$ $r_1(side)=0.304$ $r_2(side)=0.227$

$P=1^d.3057$ $r_1(back)=0.306$ $r_2(back)=0.259$

$i=86°.5$ $\Omega_1=3.496$ $V_\gamma= \text{---}$

$T_1=7700\ K$ $\Omega_2=2.130$ $F_1=1.00$

$T_2=4400\ K$ $q=0.16$ $F_2=1.00$

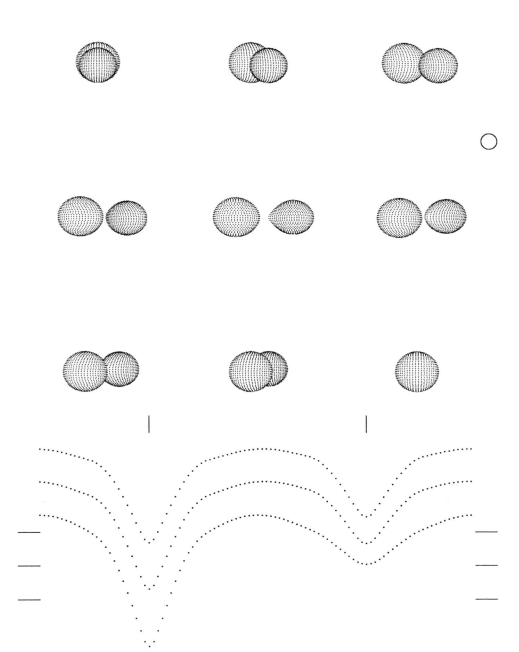

a= 7.1 R$_\odot$ r$_1$(pole)=0.363   r$_2$(pole)=0.315

e= 0.000   r$_1$(point)=0.417   r$_2$(point)=0.448

ω= −−−   r$_1$(side)=0.379   r$_2$(side)=0.329

P=1^d.0362   r$_1$(back)=0.396   r$_2$(back)=0.361

i=87°.0 Ω_1=3.319   V$_\gamma$= −−−

T$_1$=12000 K Ω_2=3.069   F$_1$=1.00

T$_2$=9170 K   q=0.603   F$_2$=1.00

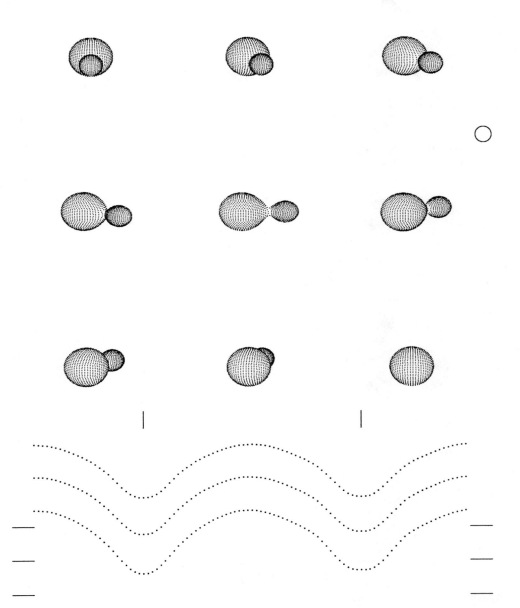

a=5.3 R$_\odot$ r$_1$(pole)=0.473   r$_2$(pole)=0.265

e= 0.000   r$_1$(point)=−1.000   r$_2$(point)=−1.000

ω= −−−   r$_1$(side)=0.512   r$_2$(side)=0.278

P=1^d.1494   r$_1$(back)=0.540   r$_2$(back)=0.320

i=78°.49 Ω_1=2.364   V$_\gamma$= −−−

T$_1$=6605 K Ω_2=2.364   F$_1$=1.00

T$_2$=6569 K   q=0.275   F$_2$=1.00

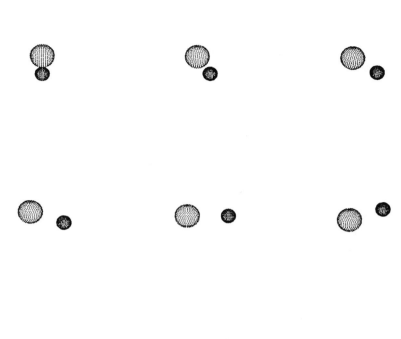

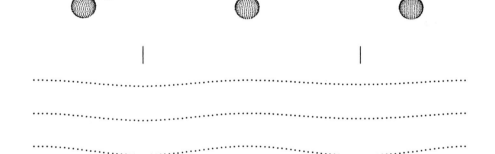

a=5.12 R$_\odot$ r$_1$(pole)=0.288   r$_2$(pole)=0.175

e= 0.000   r$_1$(point)=0.303   r$_2$(point)=0.179

ω= ---   r$_1$(side)=0.294   r$_2$(side)=0.176

P=0^d.9172   r$_1$(back)=0.299   r$_2$(back)=0.178

i=63°.0 Ω_1=3.986   V$_\gamma$=+0.2 km sec^{-1}

T$_1$=8620 K Ω_2=4.307   F$_1$=1.00

T$_2$=5920 K   q=0.54   F$_2$=1.00

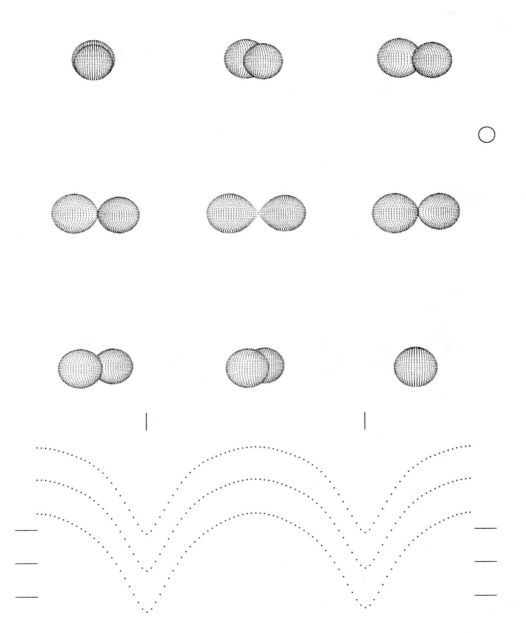

a=6.73 R$_\odot$　　　　r$_1$(pole)=0.374　　　r$_2$(pole)=0.340

e= 0.000　　　　　r$_1$(point)=−1.000　　r$_2$(point)=−1.000

ω= −−−　　　　　r$_1$(side)=0.394　　　r$_2$(side)=0.357

P=0^d.6563　　　　r$_1$(back)=0.424　　　r$_2$(back)=0.389

i=87°.06　　　　　Ω_1=3.452　　　　V$_\gamma$=+22.0 km sec^{-1}

T$_1$=17000 K　　　　Ω_2=3.452　　　　F$_1$=1.00

T$_2$=16810 K　　　　q=0.82　　　　　F$_2$=1.00

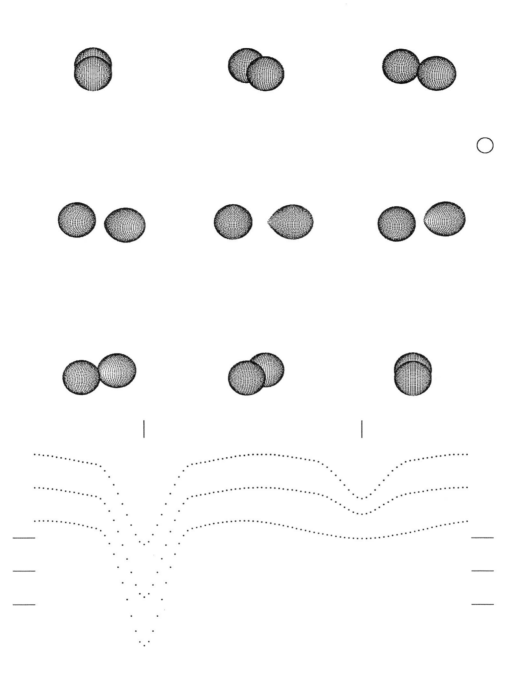

a= 7.37 $R_\odot$ r_1(pole)=0.303 r_2(pole)=0.300

e= 0.000 r_1(point)=0.320 r_2(point)=0.429

ω= --- r_1(side)=0.310 r_2(side)=0.313

P=1^d.1829 r_1(back)=0.316 r_2(back)=0.345

i=$81°$.5 Ω_1=3.775 V_γ= -2 km sec^{-1}

T_1=9150 K Ω_2=2.876 F_1=1.00

T_2=5260 K   q=0.5 F_2=1.00

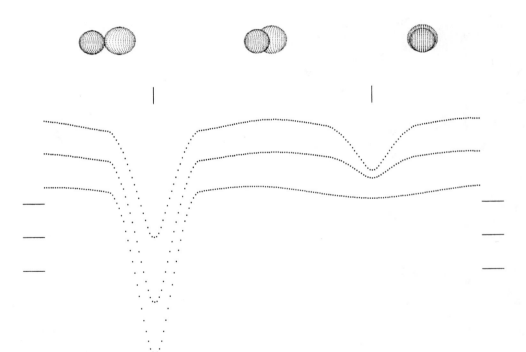

a=5.81 R$_\odot$ r$_1$(pole)=0.268   r$_2$(pole)=0.310

e= 0.000   r$_1$(point)=0.278   r$_2$(point)=0.442

ω= — — —   r$_1$(side)=0.272   r$_2$(side)=0.324

P=0^d.9715   r$_1$(back)=0.276   r$_2$(back)=0.356

i=87°.7 Ω_1=4.285   V$_\gamma$=−5.0 km sec^{-1}

T$_1$=8830 K Ω_2=3.008   F$_1$=1.00

T$_2$=4810 K   q=0.57   F$_2$=1.00

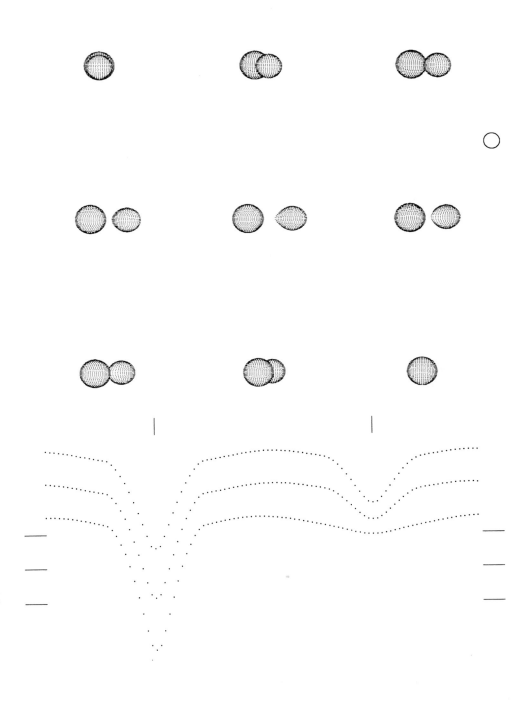

a= 5.5 $R_\odot$

e= 0.000

ω= ---

P=0^d.7131

i=$88°$.5

T_1=12000 K

T_2=7170 K

r_1(pole)=0.334

r_1(point)=0.356

r_1(side)=0.343

r_1(back)=0.350

Ω_1=3.376

Ω_2=2.678

q=0.4

r_2(pole)=0.282

r_2(point)=0.407

r_2(side)=0.294

r_2(back)=0.327

V_γ= ---

F_1=1.00

F_2=1.00

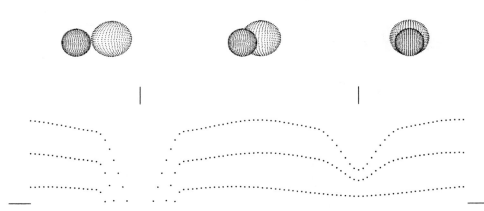

$a=7.4\ R_\odot$ $r_1(\text{pole})=0.240$ $r_2(\text{pole})=0.310$

$e=\ 0.000$ $r_1(\text{point})=0.247$ $r_2(\text{point})=0.443$

$\omega=\ ---$ $r_1(\text{side})=0.243$ $r_2(\text{side})=0.324$

$P=1^d.3155$ $r_1(\text{back})=0.246$ $r_2(\text{back})=0.357$

$i=85°.5$ $\Omega_1=4.718$ $V_\gamma=-23.8\ \text{km sec}^{-1}$

$T_1=8940\ K$ $\Omega_2=3.012$ $F_1=1.00$

$T_2=5010\ K$ $q=0.57$ $F_2=1.00$

Group V
Systems with
$7.60\ R_\odot < a \leq 11.39\ R_\odot$

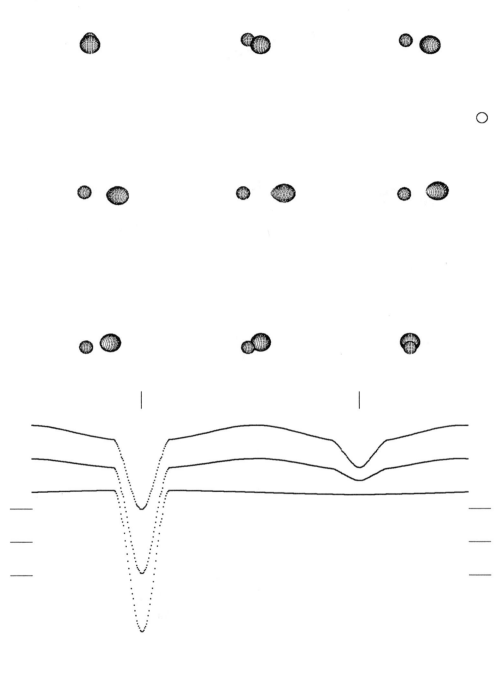

$$a= 7.61\ R_\odot \qquad r_1(pole)=0.166 \qquad r_2(pole)=0.233$$
$$e= 0.000 \qquad r_1(point)=0.167 \qquad r_2(point)=0.329$$
$$\omega= \text{---} \qquad r_1(side)=0.167 \qquad r_2(side)=0.242$$
$$P=1^d.9666 \qquad r_1(back)=0.167 \qquad r_2(back)=0.275$$
$$i=82^\circ.11 \qquad \Omega_1=6.203 \qquad V_\gamma= \text{---}$$
$$T_1=7605\ K \qquad \Omega_2=2.234 \qquad F_1=1.00$$
$$T_2=4520\ K \qquad q=0.201 \qquad F_2=1.00$$

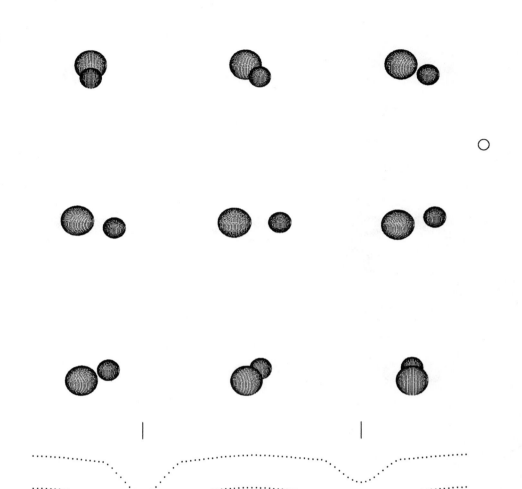

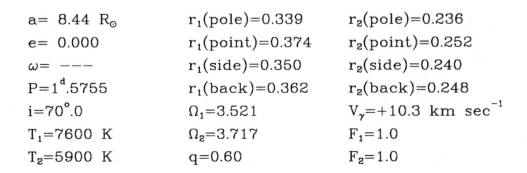

a= 8.44 $R_\odot$ r_1(pole)=0.339 r_2(pole)=0.236

e= 0.000 r_1(point)=0.374 r_2(point)=0.252

ω= – – – r_1(side)=0.350 r_2(side)=0.240

P=1^d.5755 r_1(back)=0.362 r_2(back)=0.248

i=70°.0 Ω_1=3.521 V_γ=+10.3 km sec^{-1}

T_1=7600 K Ω_2=3.717 F_1=1.0

T_2=5900 K   q=0.60 F_2=1.0

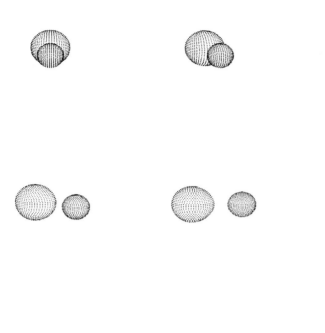

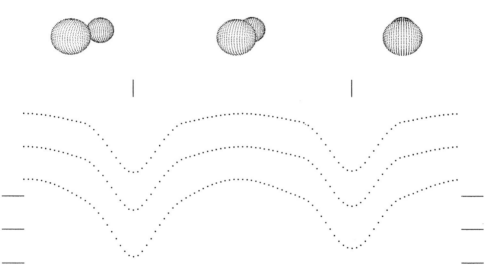

a=9.2 R$_\odot$ r$_1$(pole)=0.376   r$_2$(pole)=0.261

e= 0.000   r$_1$(point)=0.446   r$_2$(point)=0.288

ω= ———   r$_1$(side)=0.394   r$_2$(side)=0.268

P=1^d.3959   r$_1$(back)=0.415   r$_2$(back)=0.280

i=80°.46 Ω_1=3.221   V$_\gamma$= ———

T$_1$=9600 K Ω_2=3.472   F$_1$=1.00

T$_2$=9131 K   q=0.602   F$_2$=1.00

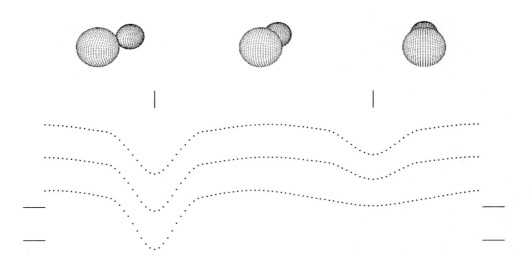

a=10.1 $R_\odot$ r_1(pole)=0.379 r_2(pole)=0.238

e= 0.000 r_1(point)=0.406 r_2(point)=0.345

ω= --- r_1(side)=0.393 r_2(side)=0.248

P=$0^d.7713$ r_1(back)=0.399 r_2(back)=0.280

i=$75^\circ.56$ Ω_1=2.842 V_γ=+3.0 km sec^{-1}

T_1=9900 K Ω_2=2.275 F_1=1.00

T_2=6381 K   q=0.217 F_2=1.00

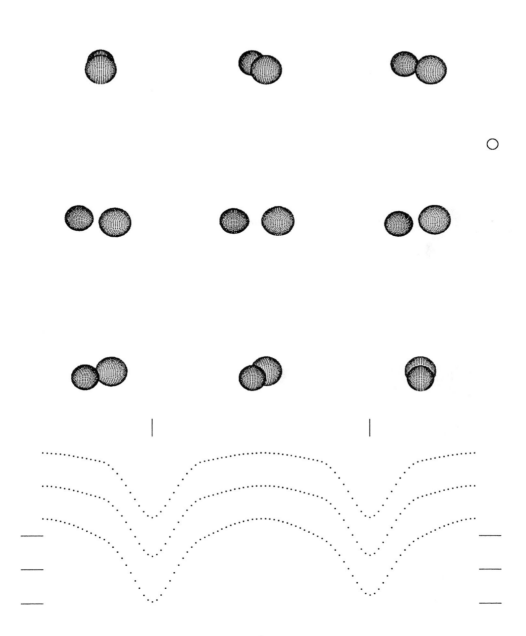

a=8.2 R$_\odot$ r$_1$(pole)=0.294   r$_2$(pole)=0.339

e= 0.000   r$_1$(point)=0.341   r$_2$(point)=0.378

ω= ---   r$_1$(side)=0.304   r$_2$(side)=0.351

P=1^d.3192   r$_1$(back)=0.324   r$_2$(back)=0.365

i=80°.51 Ω_1=4.863   V$_\gamma$= ---

T$_1$=7520 K Ω_2=5.172   F$_1$=1.00

T$_2$=7369 K   q=1.523   F$_2$=1.00

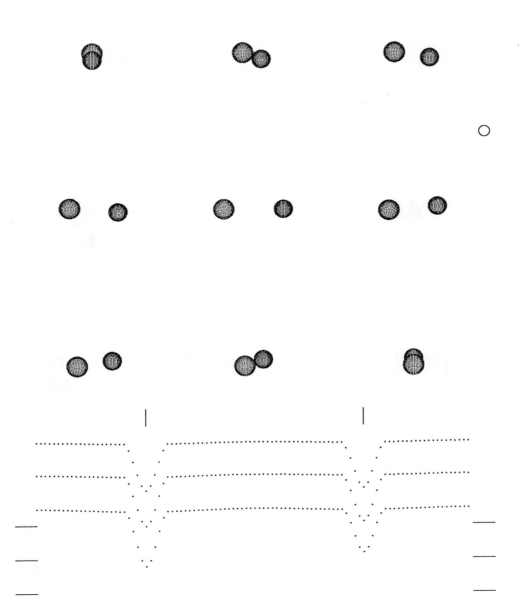

a=11.06 $R_\odot$ r_1(pole)=0.174 r_2(pole)=0.155

e= 0.000 r_1(point)=0.176 r_2(point)=0.157

ω= ——— r_1(side)=0.175 r_2(side)=0.156

P=2^d.1180 r_1(back)=0.176 r_2(back)=0.157

i=83°.3 Ω_1=6.674 V_γ=16.8 km sec^{-1}

T_1=10800 K Ω_2=7.090 F_1=1.00

T_2=10300 K   q=0.944 F_2=1.00

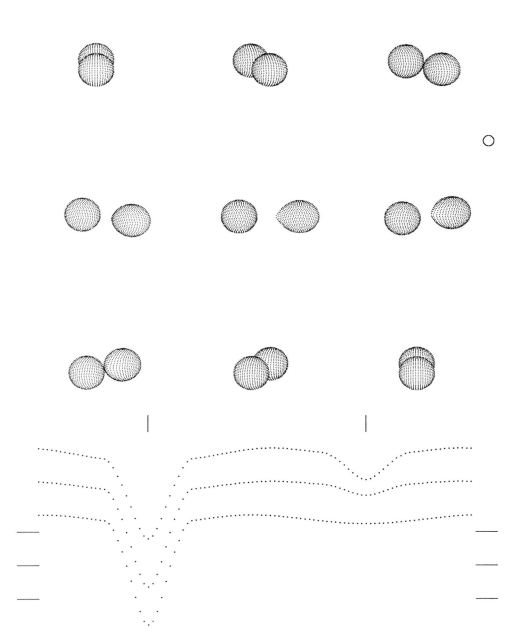

a=11.2 $R_\odot$ r_1(pole)=0.289 r_2(pole)=0.283

e= 0.000 r_1(point)=0.300 r_2(point)=0.404

ω= --- r_1(side)=0.294 r_2(side)=0.295

P=1^d.8127 r_1(back)=0.298 r_2(back)=0.327

i=80°.3 Ω_1=3.846 V_γ=+0.5 km sec^{-1}

T_1=10800 K Ω_2=2.678 F_1=1.00

T_2=5100 K   q=0.40 F_2=1.00

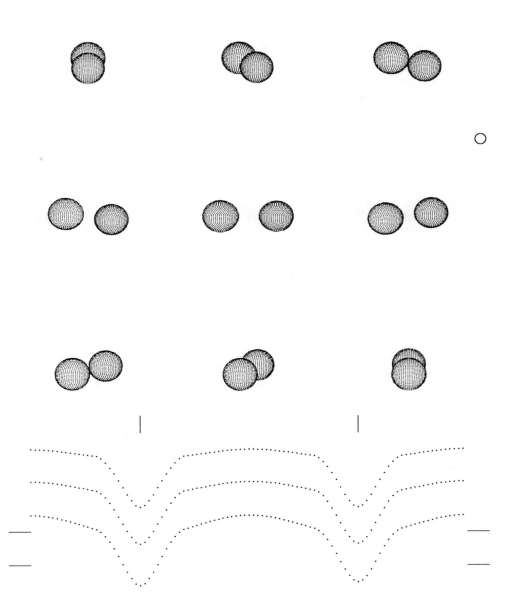

a=10.4 R$_\odot$ r$_1$(pole)=0.299   r$_2$(pole)=0.288

e= 0.000   r$_1$(point)=0.331   r$_2$(point)=0.313

ω= ---   r$_1$(side)=0.308   r$_2$(side)=0.295

P=1^d.9169   r$_1$(back)=0.321   r$_2$(back)=0.306

i=79°.4 Ω_1=4.297   V$_\gamma$= ---

T$_1$=9600 K Ω_2=4.436   F$_1$=1.00

T$_2$=9425 K   q=1.0   F$_2$=1.00

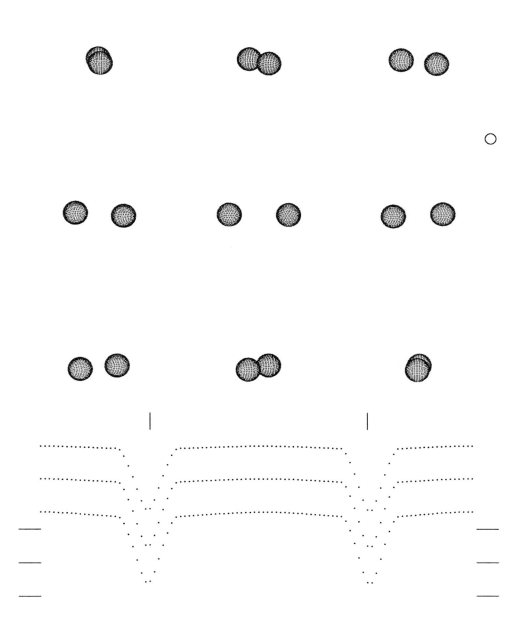

a= 10.851 $R_\odot$ r_1(pole)=0.207 r_2(pole)=0.207

e= 0.031 r_1(point)=0.213 r_2(point)=0.213

ω= 303°.1 r_1(side)=0.209 r_2(side)=0.209

P=1^d.7505 r_1(back)=0.212 r_2(back)=0.212

i=85°.35 Ω_1=5.864 V_γ= −3.3 km sec^{-1}

T_1=10350 K Ω_2=5.932 F_1=1.06

T_2=10350 K   q=1.021 F_2=1.06

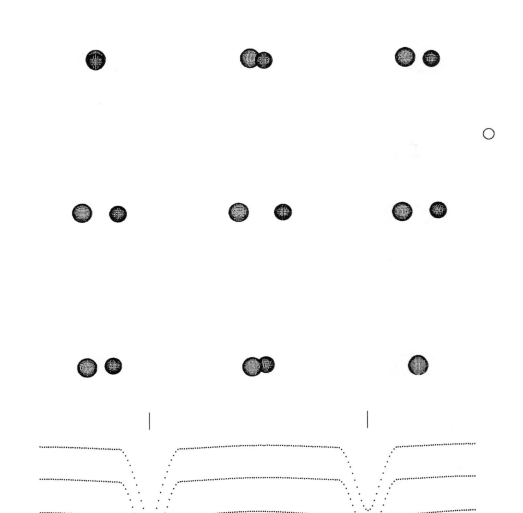

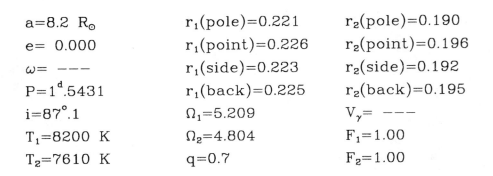

a=8.2 $R_\odot$	r_1(pole)=0.221	r_2(pole)=0.190
e= 0.000	r_1(point)=0.226	r_2(point)=0.196
ω= ---	r_1(side)=0.223	r_2(side)=0.192
P=1^d.5431	r_1(back)=0.225	r_2(back)=0.195
i=87°.1	Ω_1=5.209	V_γ= ---
T_1=8200 K	Ω_2=4.804	F_1=1.00
T_2=7610 K	q=0.7	F_2=1.00

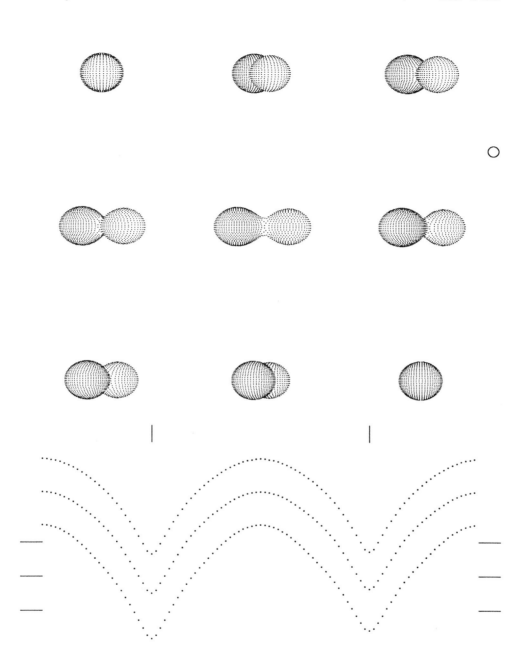

a=9.3 R$_\odot$ r$_1$(pole)=0.403   r$_2$(pole)=0.374

e= 0.000   r$_1$(point)=−1.000   r$_2$(point)=−1.000

ω= − − −   r$_1$(side)=0.431   r$_2$(side)=0.399

P=0^d.7916   r$_1$(back)=0.481   r$_2$(back)=0.454

i=90°.0 Ω_1=3.267   V$_\gamma$= − − −

T$_1$=17900 K Ω_2=3.267   F$_1$=1.00

T$_2$=17431 K   q=0.844   F$_2$=1.00

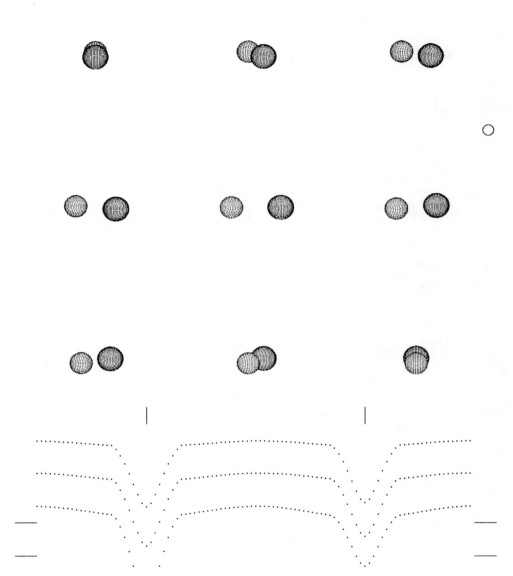

a=9.13 R$_\odot$ r$_1$(pole)=0.228   r$_2$(pole)=0.256

e= 0.000   r$_1$(point)=0.236   r$_2$(point)=0.270

ω= ---   r$_1$(side)=0.231   r$_2$(side)=0.260

P=1^d.6699   r$_1$(back)=0.235   r$_2$(back)=0.267

i=83°.9 Ω_1=5.337   V$_\gamma$=15.8 km sec^{-1}

T$_1$=7690 K Ω_2=4.814   F$_1$=1.00

T$_2$=7310 K   q=0.980   F$_2$=1.00

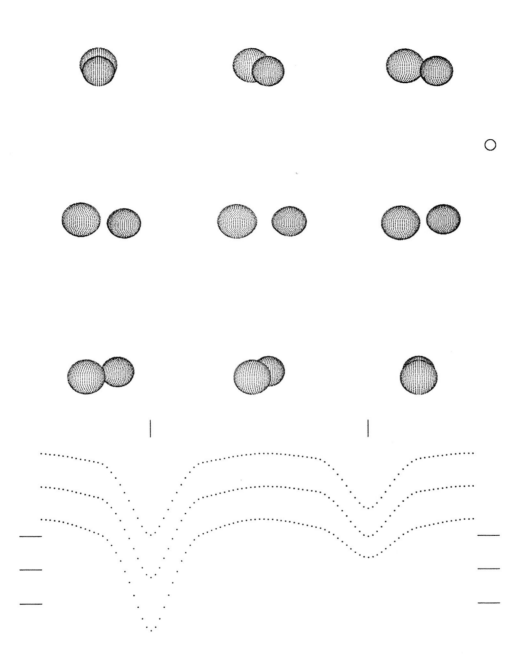

a= 9.79 $R_\odot$ r_1(pole)=0.339 r_2(pole)=0.296

e= 0.000 r_1(point)=0.382 r_2(point)=0.341

ω= --- r_1(side)=0.352 r_2(side)=0.306

P=1^d.6770 r_1(back)=0.367 r_2(back)=0.324

i=$83°$.03 Ω_1=3.611 V_γ=−54.5 km sec^{-1}

T_1=18700 K Ω_2=3.482 F_1=1.0

T_2=12934 K   q=0.702 F_2=1.0

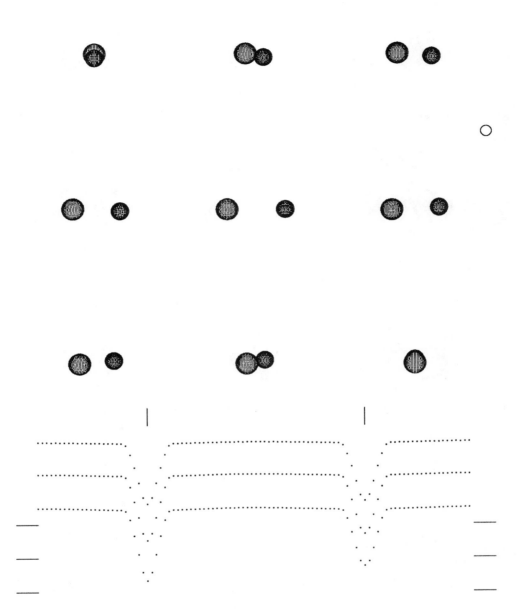

a= 10.80 $R_\odot$ r_1(pole)=0.191 r_2(pole)=0.153

e= 0.000 r_1(point)=0.194 r_2(point)=0.155

ω= ——— r_1(side)=0.192 r_2(side)=0.154

P=2^d.3859 r_1(back)=0.194 r_2(back)=0.155

i=86°.0 Ω_1=6.126 V_γ=−24.2 km sec^{-1}

T_1=6900 K Ω_2=6.904 F_1=1.0

T_2=6800 K   q=0.90 F_2=1.0

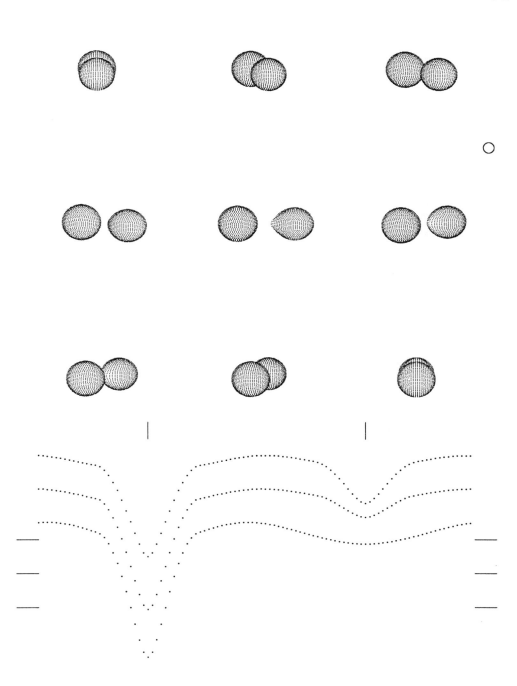

a= 10.5 R$_\odot$ r$_1$(pole)=0.327   r$_2$(pole)=0.300

e= 0.000   r$_1$(point)=0.351   r$_2$(point)=0.429

ω= ---   r$_1$(side)=0.336   r$_2$(side)=0.313

P=2^{d}.1176   r$_1$(back)=0.345   r$_2$(back)=0.345

i=84°.0 Ω_1=3.534   V$_\gamma$= ---

T$_1$=9500 K Ω_2=2.876   F$_1$=1.00

T$_2$=5280 K   q=0.5   F$_2$=1.00

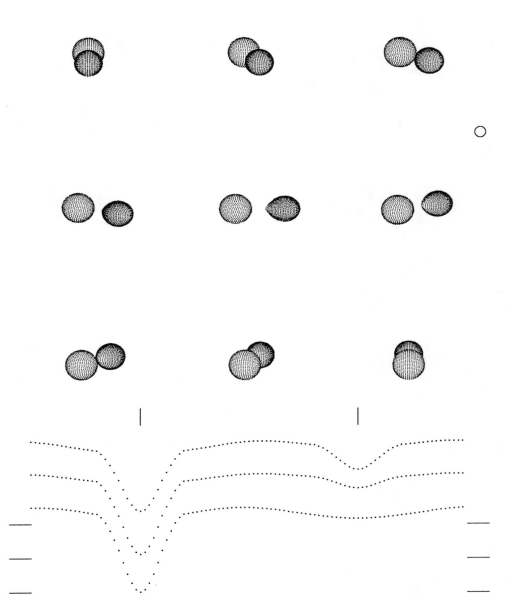

a= 9.10 $R_\odot$	r_1(pole)=0.318	r_2(pole)=0.275
e= 0.000	r_1(point)=0.334	r_2(point)=0.397
ω= ---	r_1(side)=0.325	r_2(side)=0.286
P=1^d.8052	r_1(back)=0.331	r_2(back)=0.320
i=77°.24	Ω_1=3.493	V_γ= −22.4 km sec^{-1}
T_1=10000 K	Ω_2=2.603	F_1=1.00
T_2=4914 K	q=0.364	F_2=1.00

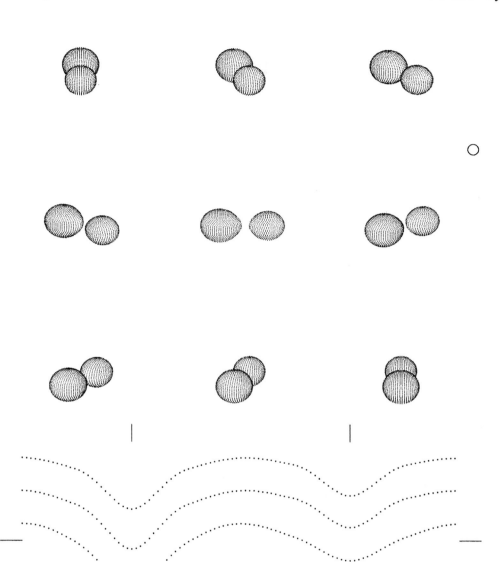

a= 10.0 R$_\odot$ r$_1$(pole)=0.368   r$_2$(pole)=0.321

e= 0.000   r$_1$(point)=0.459   r$_2$(point)=0.392

ω= ———   r$_1$(side)=0.386   r$_2$(side)=0.334

P=1^d.2524   r$_1$(back)=0.412   r$_2$(back)=0.359

i=70°.1 Ω_1=3.450   V$_\gamma$= ———

T$_1$=15500 K Ω_2=3.490   F$_1$=1.0

T$_2$=11080 K   q=0.78   F$_2$=1.0

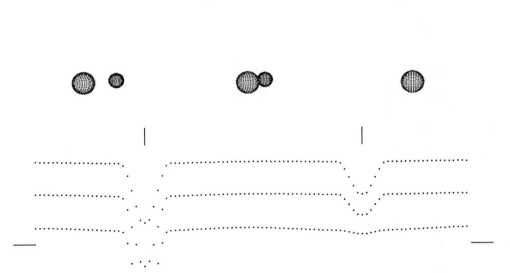

a= 10.34 R$_\odot$ r$_1$(pole)=0.205   r$_2$(pole)=0.136

e= 0.000   r$_1$(point)=0.207   r$_2$(point)=0.138

ω= ---   r$_1$(side)=0.206   r$_2$(side)=0.137

P=1^d.8172   r$_1$(back)=0.207   r$_2$(back)=0.138

i=86°.6 Ω_1=5.305   V$_\gamma$= −36.4 km sec^{-1}

T$_1$=9500 K Ω_2=4.428   F$_1$=1.00

T$_2$=6610 K   q=0.43   F$_2$=1.00

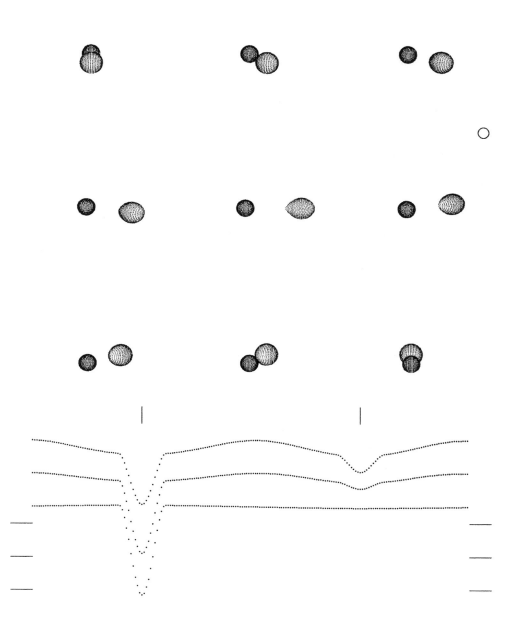

a=10.36 $R_\odot$ r_1(pole)=0.158 r_2(pole)=0.198

e= 0.000 r_1(point)=0.159 r_2(point)=0.293

ω= --- r_1(side)=0.159 r_2(side)=0.206

P=2^d.6642 r_1(back)=0.159 r_2(back)=0.237

i=80°.02 Ω_1=6.427 V_γ=+11.8 km sec^{-1}

T_1=8500 K Ω_2=2.003 F_1=1.00

T_2=4725 K   q=0.114 F_2=1.00

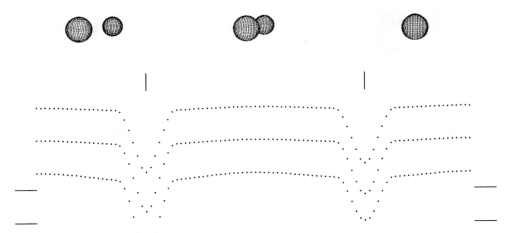

a= 10.61 $R_\odot$ r_1(pole)=0.228 r_2(pole)=0.173

e= 0.000 r_1(point)=0.236 r_2(point)=0.176

ω= --- r_1(side)=0.231 r_2(side)=0.174

P=1^d.7786 r_1(back)=0.234 r_2(back)=0.176

i=86°.3 Ω_1=5.206 V_γ=−26 km sec^{-1}

T_1=10500 K Ω_2=5.973 F_1=1.00

T_2=96600 K   q=0.85 F_2=1.00

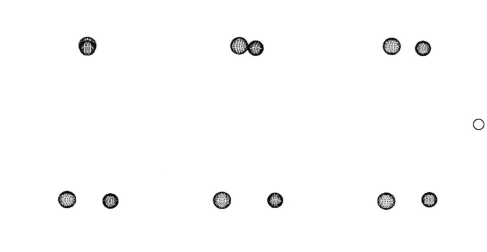

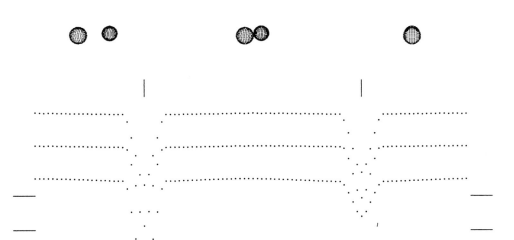

a= 9.88 R$_\odot$ r$_1$(pole)=0.164   r$_2$(pole)=0.145

e= 0.000   r$_1$(point)=0.166   r$_2$(point)=0.146

ω= ---   r$_1$(side)=0.165   r$_2$(side)=0.145

P=2^d.0598   r$_1$(back)=0.166   r$_2$(back)=0.146

i=87°.2 Ω_1=6.973   V$_\gamma$= −8.8 km sec^{-1}

T$_1$=7840 K Ω_2=7.980   F$_1$=1.00

T$_2$=6980 K   q=0.90   F$_2$=1.00

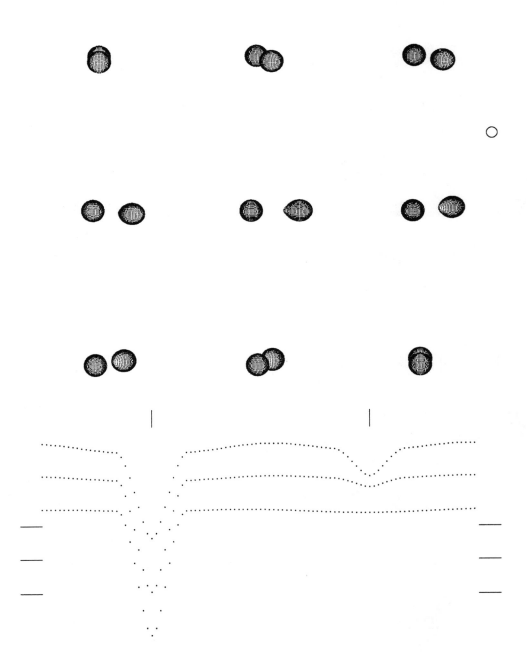

a= 8.92 $R_\odot$ r_1(pole)=0.237 r_2(pole)=0.236

e= 0.000 r_1(point)=0.240 r_2(point)=0.346

ω= ——— r_1(side)=0.239 r_2(side)=0.246

P=1^d.5489 r_1(back)=0.240 r_2(back)=0.278

i=83°.1 Ω_1=4.421 V_γ= −58.6 km sec^{-1}

T_1=8870 K Ω_2=2.257 F_1=1.00

T_2=4360 K   q=0.21 F_2=1.00

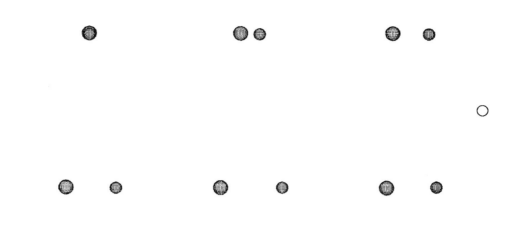

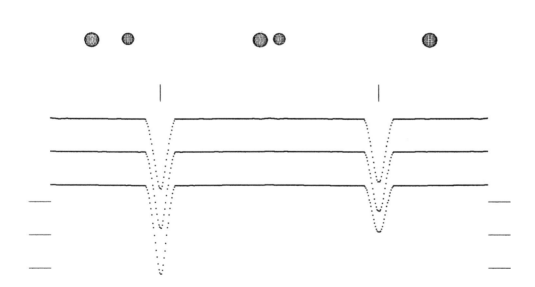

a= 11.38 $R_\odot$ r_1(pole)=0.119 r_2(pole)=0.098

e= 0.000 r_1(point)=0.120 r_2(point)=0.098

ω= − − − r_1(side)=0.120 r_2(side)=0.098

P=2^d.9043 r_1(back)=0.120 r_2(back)=0.098

i=89°.0 Ω_1=9.250 V_γ=+0.0 km sec^{-1}

T_1=6450 K Ω_2=10.335 F_1=1.00

T_2=6120 K   q=0.91 F_2=1.00

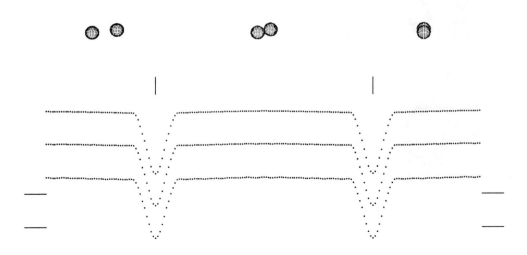

a=7.82 R$_\odot$ r$_1$(pole)=0.163   r$_2$(pole)=0.164

e= 0.000   r$_1$(point)=0.165   r$_2$(point)=0.166

ω= ---   r$_1$(side)=0.164   r$_2$(side)=0.164

P=1^d.5680   r$_1$(back)=0.165   r$_2$(back)=0.165

i=85°.35 Ω_1=7.107   V$_\gamma$=−8.7 km sec^{-1}

T$_1$=6770 K Ω_2=7.100   F$_1$=1.00

T$_2$=6770 K   q=1.0   F$_2$=1.00

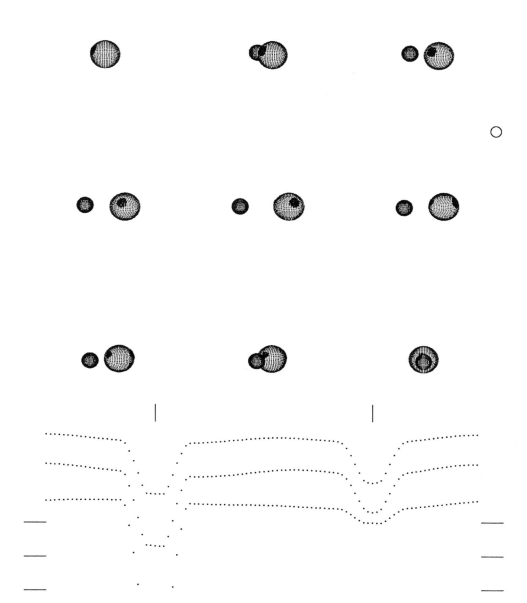

$$a=\ 9.09\ R_{\odot} \qquad r_1(pole)=0.164 \qquad r_2(pole)=0.293$$

$$e=\ 0.000 \qquad r_1(point)=0.166 \qquad r_2(point)=0.320$$

$$\omega=\ --- \qquad r_1(side)=0.165 \qquad r_2(side)=0.301$$

$$P=1^d.9832 \qquad r_1(back)=0.166 \qquad r_2(back)=0.312$$

$$i=87^{\circ}.0 \qquad \Omega_1=6.970 \qquad V_{\gamma}=-33.7\ km\ sec^{-1}$$

$$T_1=5368\ K \qquad \Omega_2=4.374 \qquad F_1=1.0$$

$$T_2=4550\ K \qquad q=1.00 \qquad F_2=1.0$$

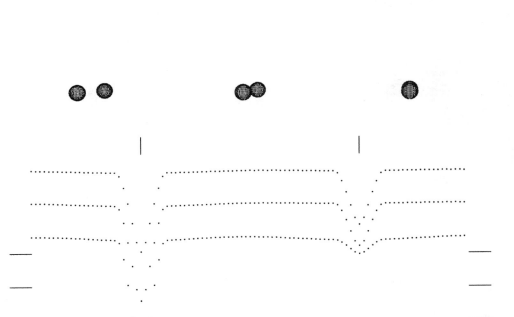

a= 8.63 R$_\odot$ r$_1$(pole)=0.178   r$_2$(pole)=0.170

e= 0.000   r$_1$(point)=0.180   r$_2$(point)=0.173

ω= ———   r$_1$(side)=0.179   r$_2$(side)=0.171

P=1^d.6047   r$_1$(back)=0.180   r$_2$(back)=0.173

i=86°.8 Ω_1=6.380   V$_\gamma$=−16.1km sec^{-1}

T$_1$=8800 K Ω_2=5.681   F$_1$=1.00

T$_2$=7000 K   q=0.78   F$_2$=1.00

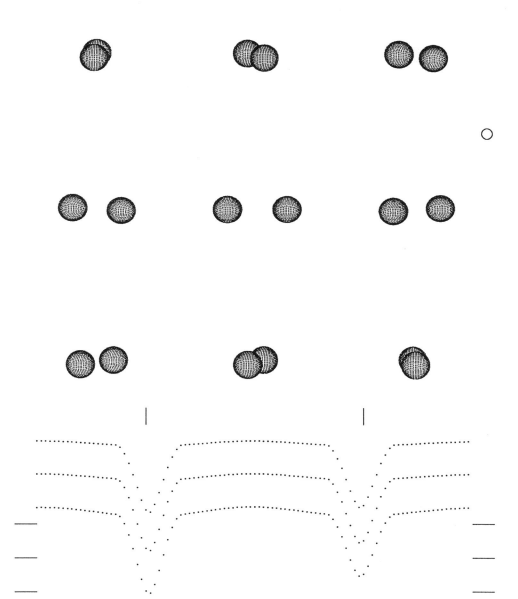

a= 11.33 $R_\odot$ r_1(pole)=0.232 r_2(pole)=0.227

e= 0.032 r_1(point)=0.240 r_2(point)=0.238

ω= 132°.5 r_1(side)=0.234 r_2(side)=0.231

P=1^d.5422 r_1(back)=0.238 r_2(back)=0.236

i=85°.26 Ω_1=5.143 V_γ= −32.5 km sec^{-1}

T_1=11350 K Ω_2=4.704 F_1=1.06

T_2=10952 K   q=0.82 F_2=1.06

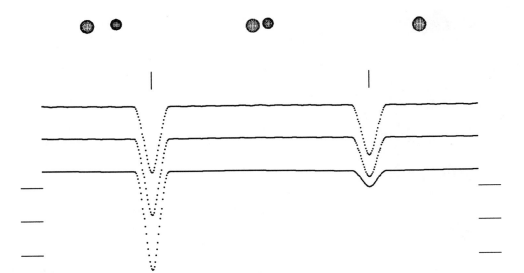

a= 9.14 $R_\odot$	r_1(pole)=0.138	r_2(pole)=0.112
e= 0.000	r_1(point)=0.138	r_2(point)=0.112
ω= ---	r_1(side)=0.138	r_2(side)=0.112
P=2^d.1782	r_1(back)=0.138	r_2(back)=0.112
i=$87°$.2	Ω_1=8.046	V_γ=−39 km sec^{-1}
T_1=6000 K	Ω_2=8.160	F_1=1.00
T_2=5230 K	q=0.79	F_2=1.00

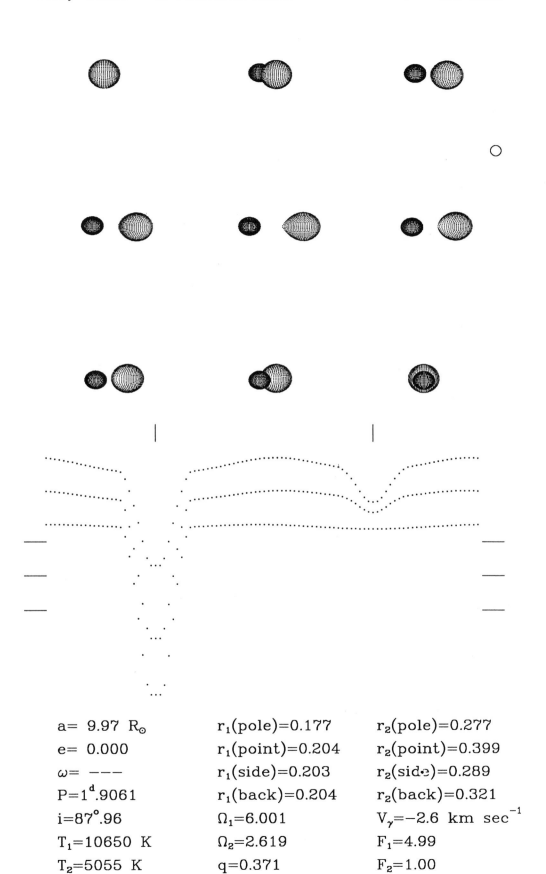

$a= 9.97\ R_\odot$ $r_1(\text{pole})=0.177$ $r_2(\text{pole})=0.277$

$e= 0.000$ $r_1(\text{point})=0.204$ $r_2(\text{point})=0.399$

$\omega= \text{---}$ $r_1(\text{side})=0.203$ $r_2(\text{side})=0.289$

$P=1^{d}.9061$ $r_1(\text{back})=0.204$ $r_2(\text{back})=0.321$

$i=87^\circ.96$ $\Omega_1=6.001$ $V_\gamma=-2.6\ \text{km sec}^{-1}$

$T_1=10650\ K$ $\Omega_2=2.619$ $F_1=4.99$

$T_2=5055\ K$ $q=0.371$ $F_2=1.00$

a= 11.31 $R_\odot$	r_1(pole)=0.322	r_2(pole)=0.183
e= 0.000	r_1(point)=0.352	r_2(point)=0.188
ω= ---	r_1(side)=0.332	r_2(side)=0.185
P=1^d.1903	r_1(back)=0.344	r_2(back)=0.188
i=65°.1	Ω_1=3.730	V_γ=8.3 km sec^{-1}
T_1=15110 K	Ω_2=4.749	F_1=1.00
T_2=11150 K	q=0.660	F_2=1.00

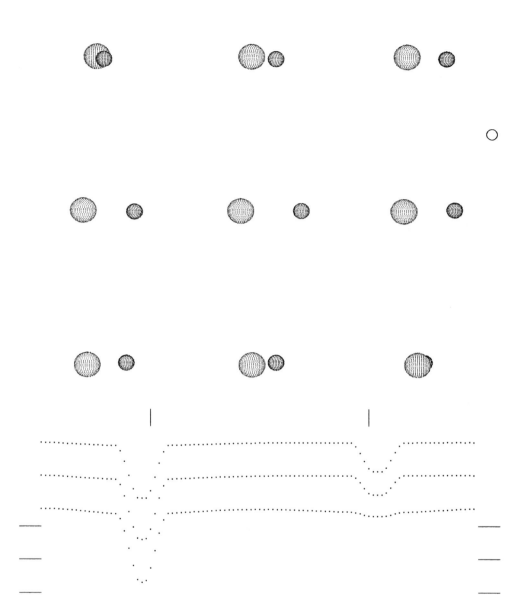

a=10.58 $R_\odot$ r_1(pole)=0.228 r_2(pole)=0.141

e=0.061 r_1(point)=0.234 r_2(point)=0.144

ω=8°.43 r_1(side)=0.231 r_2(side)=0.142

P=1^d.7436 r_1(back)=0.233 r_2(back)=0.143

i=87°.4 Ω_1=4.890 V_γ=+1.7 km sec^{-1}

T_1=12530 K Ω_2=4.798 F_1=1.13

T_2=7890 K   q=0.493 F_2=1.13

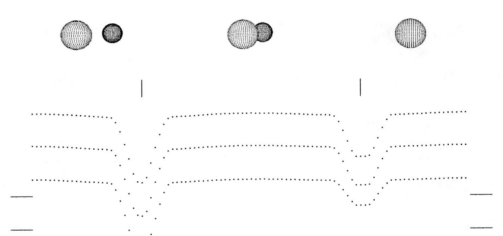

a=11.02 $R_\odot$ r_1(pole)=0.255 r_2(pole)=0.167

e=0.011 r_1(point)=0.264 r_2(point)=0.170

ω=0°.0 r_1(side)=0.258 r_2(side)=0.168

P=1^d.6698 r_1(back)=0.262 r_2(back)=0.170

i=87°.8 Ω_1=4.564 V_γ=+14.4 km sec^{-1}

T_1=14000 K Ω_2=5.064 F_1=1.02

T_2=11698 K   q=0.65 F_2=1.02

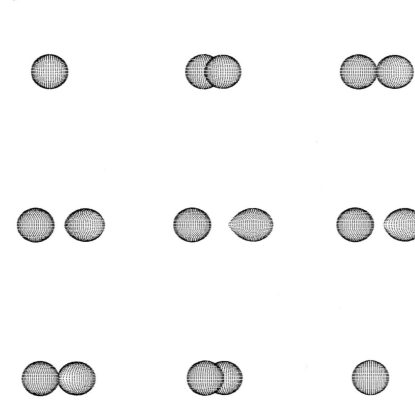

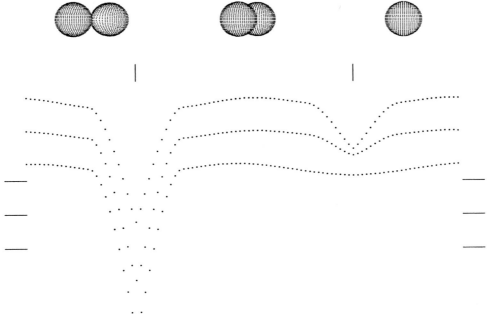

a=11.27 $R_\odot$	r_1(pole)=0.295	r_2(pole)=0.286
e= 0.000	r_1(point)=0.308	r_2(point)=0.412
ω= ---	r_1(side)=0.301	r_2(side)=0.298
P=2^d.1924	r_1(back)=0.306	r_2(back)=0.331
i=90°.0	Ω_1=3.789	V_γ=21.4 km sec^{-1}
T_1=9600 K	Ω_2=2.719	F_1=1.00
T_2=5120 K	q=0.420	F_2=1.00

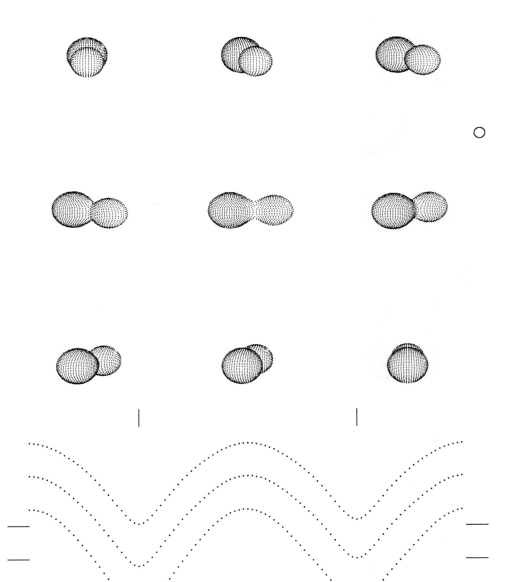

a=7.9 $R_{\odot}$ r_1(pole)=0.437 r_2(pole)=0.366

e= 0.000 r_1(point)=−1.000 r_2(point)=−1.000

ω= --- r_1(side)=0.473 r_2(side)=0.392

P=1^d.1264 r_1(back)=0.531 r_2(back)=0.472

i=80°.99 Ω_1=2.876 V_{γ}= ---

T_1=9600 K Ω_2=2.876 F_1=1.00

T_2=9067 K   q=0.644 F_2=1.00

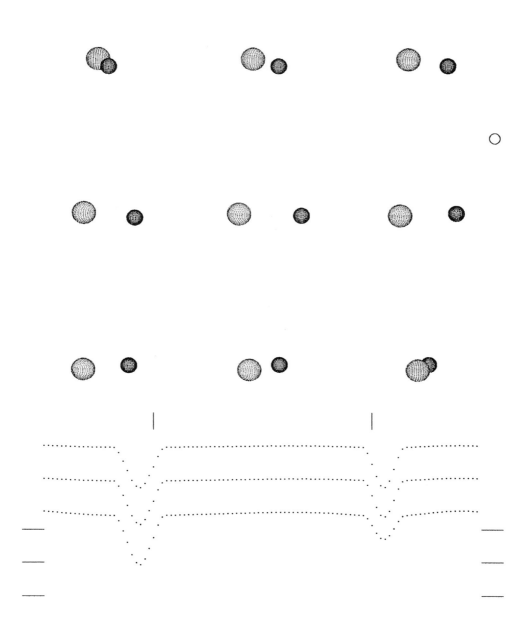

a=10.7 R$_\odot$ r$_1$(pole)=0.202   r$_2$(pole)=0.139

e=0.153   r$_1$(point)=0.209   r$_2$(point)=0.142

ω=307°.0   r$_1$(side)=0.205   r$_2$(side)=0.140

P=2^d.1468   r$_1$(back)=0.208   r$_2$(back)=0.141

i=83°.0 Ω_1=5.749   V$_\gamma$= ---

T$_1$=9800 K Ω_2=6.346   F$_1$=1.38

T$_2$=8500 K   q=0.7   F$_2$=1.38

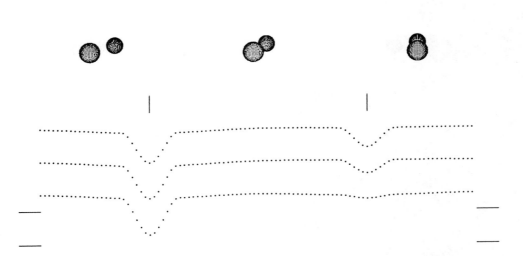

a=7.9 R$_\odot$ r$_1$(pole)=0.241   r$_2$(pole)=0.193

e= 0.000   r$_1$(point)=0.247   r$_2$(point)=0.201

ω= –––   r$_1$(side)=0.244   r$_2$(side)=0.195

P=1^d.2569   r$_1$(back)=0.246   r$_2$(back)=0.199

i=77°.8 Ω_1=4.629   V$_\gamma$= –––

T$_1$=12000 K Ω_2=3.824   F$_1$=1.00

T$_2$=7900 K   q=0.5   F$_2$=1.00

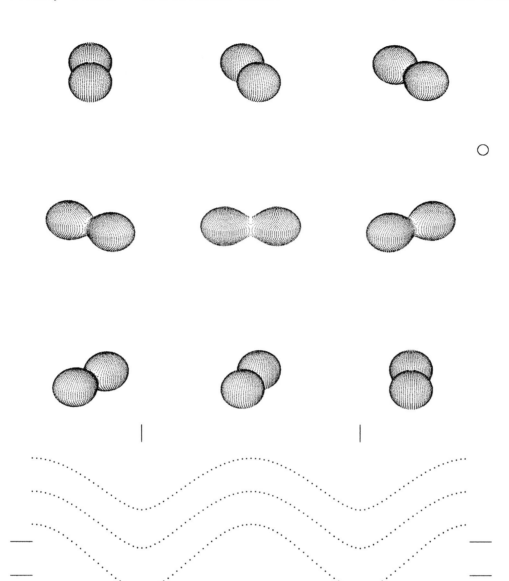

a= 9.48 $R_\odot$ r_1(pole)=0.393 r_2(pole)=0.393

e= 0.000 r_1(point)=−1.000 r_2(point)=−1.000

ω= −−− r_1(side)=0.420 r_2(side)=0.420

P=0^d.7619 r_1(back)=0.479 r_2(back)=0.479

i=66°.76 Ω_1=3.471 V_γ= +7 km sec^{-1}

T_1=20500 K Ω_2=3.471 F_1=1.0

T_2=20586 K   q=1.0 F_2=1.0

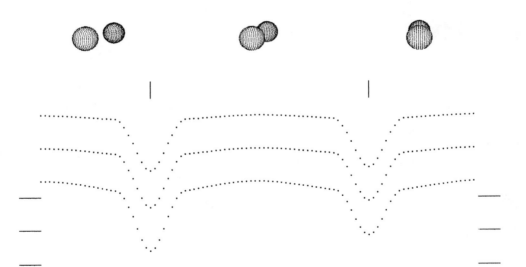

a= 9.0 R$_\odot$	r$_1$(pole)=0.261	r$_2$(pole)=0.220
e= 0.000	r$_1$(point)=0.274	r$_2$(point)=0.227
ω= ---	r$_1$(side)=0.265	r$_2$(side)=0.222
P=1$^{\rm d}$.6131	r$_1$(back)=0.271	r$_2$(back)=0.226
i=81°.5	Ω_1=4.708	V$_\gamma$= ---
T$_1$=9000 K	Ω_2=5.127	F$_1$=1.00
T$_2$=8580 K	q=0.9	F$_2$=1.00

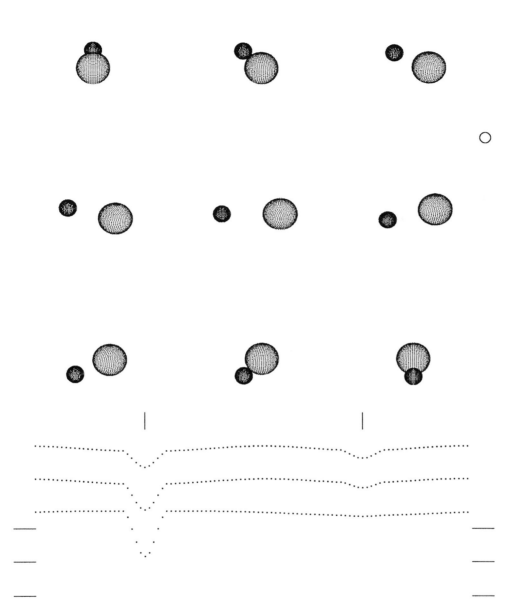

a= 10.85 R$_\odot$ r$_1$(pole)=0.149   r$_2$(pole)=0.281

e= 0.000   r$_1$(point)=0.151   r$_2$(point)=0.300

ω= ———   r$_1$(side)=0.150   r$_2$(side)=0.287

P=2^d.7977   r$_1$(back)=0.151   r$_2$(back)=0.295

i=71°.5 Ω_1=7.818   V$_\gamma$= +12.1 km sec^{-1}

T$_1$=6100 K Ω_2=4.952   F$_1$=1.00

T$_2$=4500 K   q=1.14   F$_2$=1.00

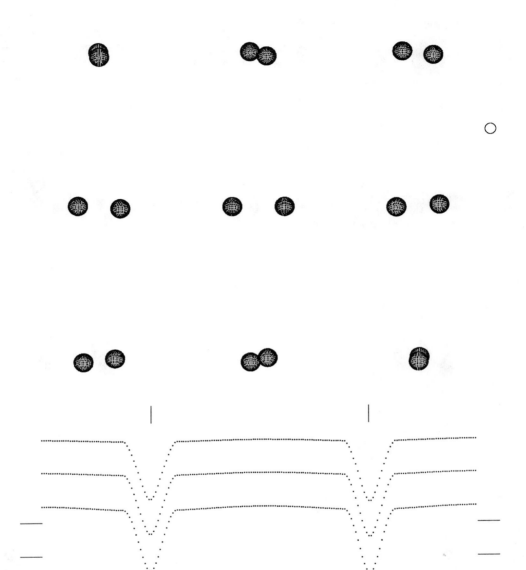

a=9.64 R☉ r_1(pole)=0.186 r_2(pole)=0.186

e=0.015 r_1(point)=0.190 r_2(point)=0.190

ω=302°.50 r_1(side)=0.188 r_2(side)=0.188

P=1^d.7304 r_1(back)=0.190 r_2(back)=0.189

i=84°.83 Ω_1=6.420 V_γ=−1.6 km sec^{-1}

T_1=7700 K Ω_2=6.638 F_1=1.03

T_2=7700 K   q=1.057 F_2=1.03

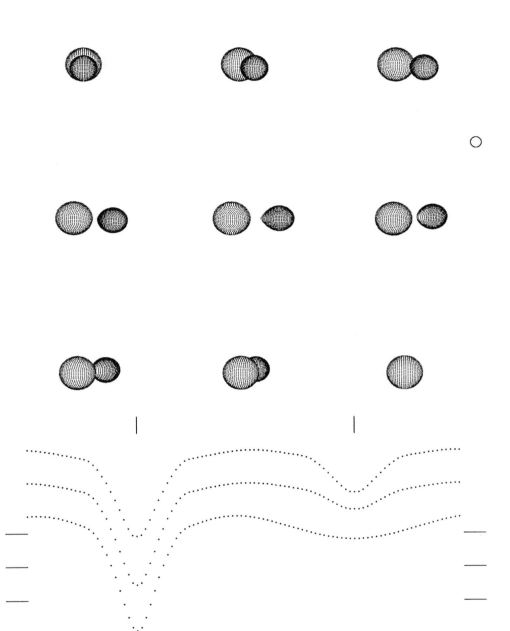

a= 11.00 R$_\odot$ r$_1$(pole)=0.365   r$_2$(pole)=0.274

e= 0.000   r$_1$(point)=0.397   r$_2$(point)=0.397

ω= ---   r$_1$(side)=0.378   r$_2$(side)=0.286

P=1^{d}.7012   r$_1$(back)=0.388   r$_2$(back)=0.319

i=85°.4 Ω_1=3.079   V$_\gamma$=−17.9 km sec^{-1}

T$_1$=9000 K Ω_2=2.595   F$_1$=1.00

T$_2$=5240 K   q=0.36   F$_2$=1.00

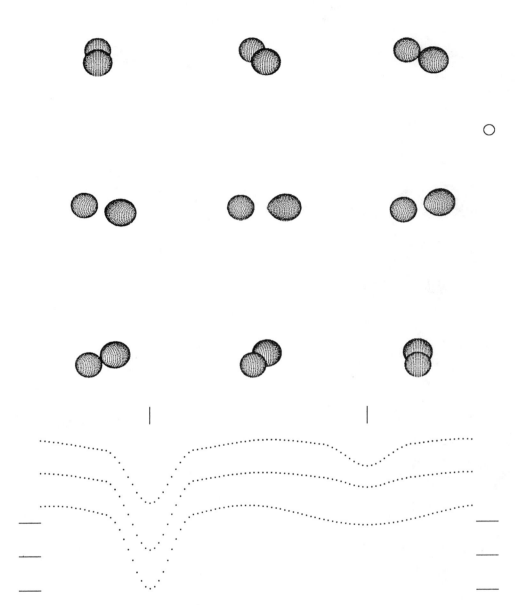

a= 9.1 $R_\odot$ r_1(pole)=0.289 r_2(pole)=0.312

e= 0.000 r_1(point)=0.305 r_2(point)=0.408

ω= ——— r_1(side)=0.295 r_2(side)=0.326

P=1^d.8420 r_1(back)=0.301 r_2(back)=0.356

i=73°.33 Ω_1=4.051 V_γ= ———

T_1=7100 K Ω_2=3.118 F_1=1.00

T_2=3940 K   q=0.616 F_2=1.00

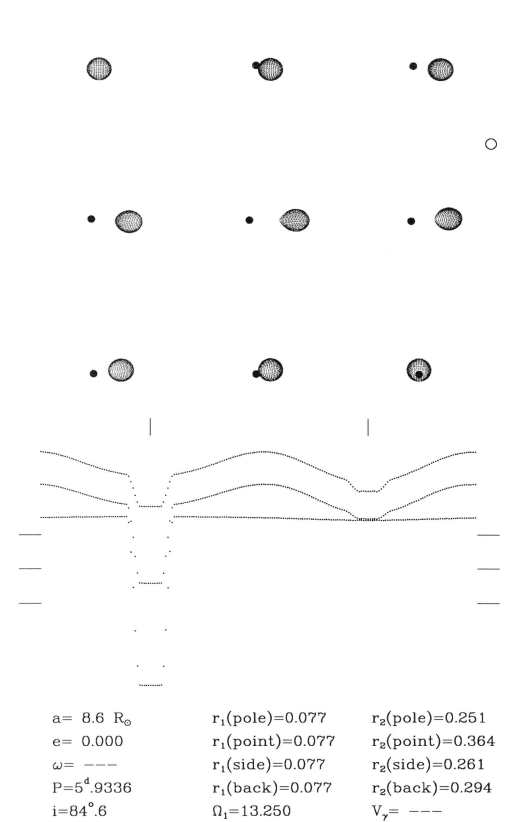

a= 8.6 R$_\odot$	r$_1$(pole)=0.077	r$_2$(pole)=0.251
e= 0.000	r$_1$(point)=0.077	r$_2$(point)=0.364
ω= ---	r$_1$(side)=0.077	r$_2$(side)=0.261
P=5^d.9336	r$_1$(back)=0.077	r$_2$(back)=0.294
i=84°.6	Ω_1=13.250	V$_\gamma$= ---
T$_1$=8200 K	Ω_2=2.376	F$_1$=1.00
T$_2$=4500 K	q=0.26	F$_2$=1.00

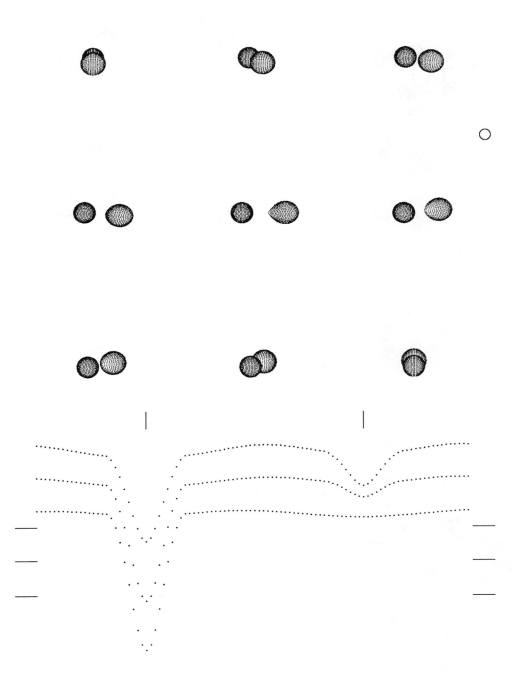

a= 8.0 R$_\odot$ r$_1$(pole)=0.252   r$_2$(pole)=0.277

e= 0.000   r$_1$(point)=0.258   r$_2$(point)=0.399

ω= ———   r$_1$(side)=0.255   r$_2$(side)=0.288

P=1^d.5520   r$_1$(back)=0.257   r$_2$(back)=0.321

i=82°.8 Ω_1=4.324   V$_\gamma$= ———

T$_1$=9840 K Ω_2=2.616   F$_1$=1.00

T$_2$=5160 K   q=0.37   F$_2$=1.00

Group VI

Systems with
$11.39\ R_\odot < a \leqslant 17.08\ R_\odot$

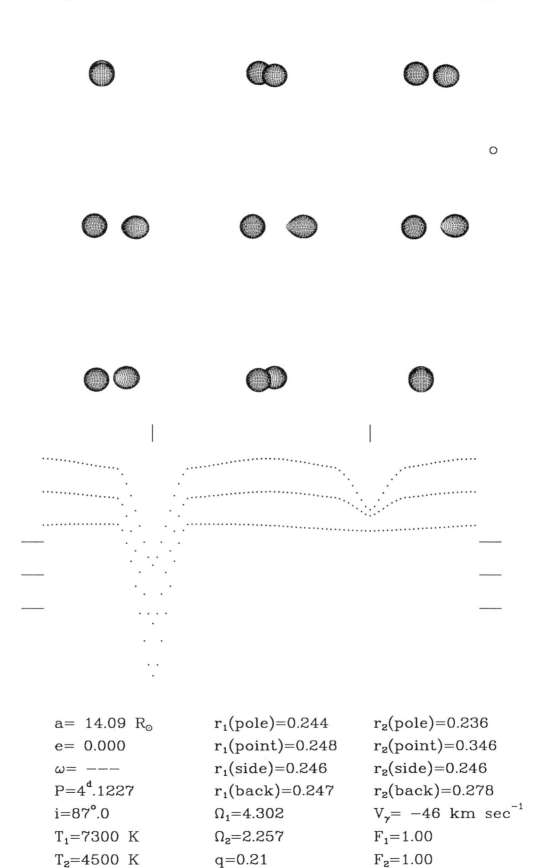

a= 14.09 $R_\odot$ r_1(pole)=0.244 r_2(pole)=0.236

e= 0.000 r_1(point)=0.248 r_2(point)=0.346

ω= ——— r_1(side)=0.246 r_2(side)=0.246

P=4^{d}.1227 r_1(back)=0.247 r_2(back)=0.278

i=87°.0 Ω_1=4.302 V_γ= −46 km sec^{-1}

T_1=7300 K Ω_2=2.257 F_1=1.00

T_2=4500 K   q=0.21 F_2=1.00

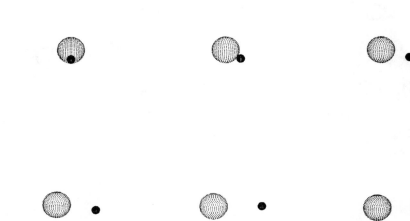

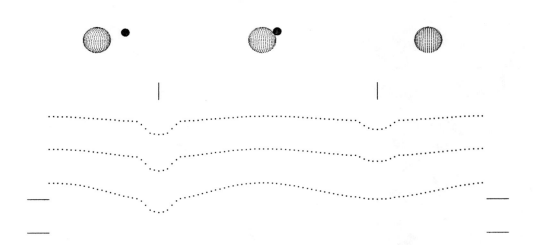

a= 12.39 $R_\odot$	$r_1(pole)=0.279$	$r_2(pole)=0.083$
e= 0.000	$r_1(point)=0.292$	$r_2(point)=0.083$
$\omega=$ ———	$r_1(side)=0.284$	$r_2(side)=0.083$
$P=3^d.2196$	$r_1(back)=0.289$	$r_2(back)=0.083$
$i=78°.6$	$\Omega_1=4.160$	$V_\gamma= -3.8$ km sec^{-1}
$T_1=7800$ K	$\Omega_2=8.404$	$F_1=1.00$
$T_2=5960$ K	$q=0.6$	$F_2=1.00$

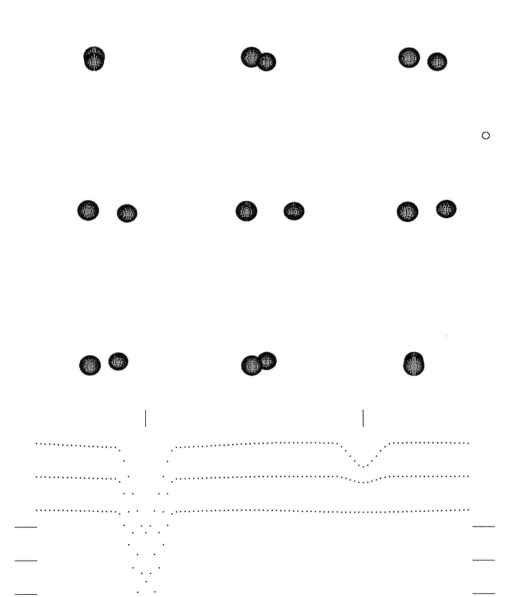

$$a= 13.30 \ R_\odot \qquad r_1(pole)=0.217 \qquad r_2(pole)=0.193$$

e= 0.000 r_1(point)=0.219 r_2(point)=0.217

ω= --- r_1(side)=0.218 r_2(side)=0.198

P=2^d.8638 r_1(back)=0.218 r_2(back)=0.211

i=83°.9 Ω_1=4.801 V_γ= −2.6 km sec^{-1}

T_1=9500 K Ω_2=2.369 F_1=1.00

T_2=4290 K   q=0.19 F_2=1.00

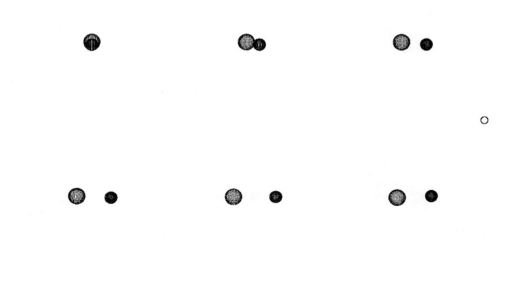

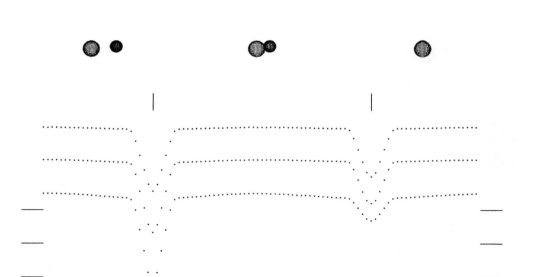

a= 11.70 R$_\odot$ r$_1$(pole)=0.198   r$_2$(pole)=0.146

e= 0.000   r$_1$(point)=0.201   r$_2$(point)=0.148

ω= – – –   r$_1$(side)=0.199   r$_2$(side)=0.147

P=2^d.4082   r$_1$(back)=0.201   r$_2$(back)=0.148

i=86°.2 Ω_1=5.815   V$_\gamma$=−36.8 km sec^{-1}

T$_1$=8990 K Ω_2=6.362   F$_1$=1.00

T$_2$=7710 K   q=0.77   F$_2$=1.00

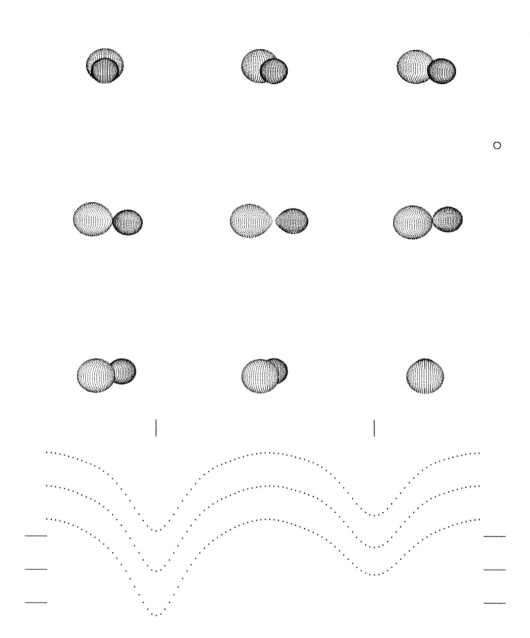

$$a=11.99 \ R_\odot \qquad r_1(pole)=0.406 \qquad r_2(pole)=0.306$$

$$e= \ 0.000 \qquad r_1(point)=0.537 \qquad r_2(point)=0.425$$

$$\omega= \ --- \qquad r_1(side)=0.430 \qquad r_2(side)=0.319$$

$$P=1^d.2101 \qquad r_1(back)=0.458 \qquad r_2(back)=0.352$$

$$i=82°.82 \qquad \Omega_1=2.964 \qquad V_\gamma=+5.0 \ km \ sec^{-1}$$

$$T_1=22000 \ K \qquad \Omega_2=2.958 \qquad F_1=1.00$$

$$T_2=16900 \ K \qquad q=0.542 \qquad F_2=1.00$$

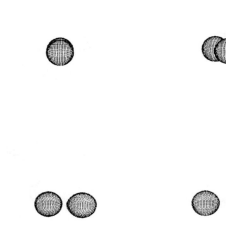

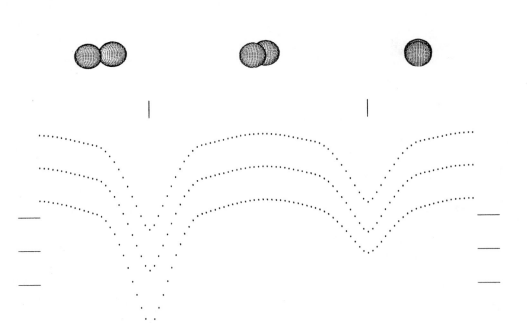

a= 11.46 $R_\odot$ r_1(pole)=0.316 r_2(pole)=0.318

e= 0.000 r_1(point)=0.342 r_2(point)=0.452

ω= ——— r_1(side)=0.325 r_2(side)=0.332

P=1^d.3327 r_1(back)=0.335 r_2(back)=0.364

i=86°.54 Ω_1=3.759 V_γ=−6.1 km sec^{-1}

T_1=24800 K Ω_2=3.115 F_1=1.00

T_2=18205 K   q=0.628 F_2=1.00

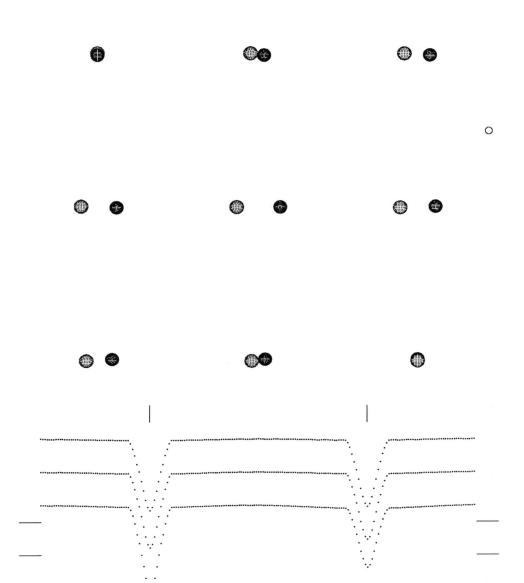

$$a= 12.14 \ R_\odot \qquad r_1(pole)=0.161 \qquad r_2(pole)=0.154$$

$$e= 0.000 \qquad r_1(point)=0.163 \qquad r_2(point)=0.156$$

$$\omega= \text{---} \qquad r_1(side)=0.162 \qquad r_2(side)=0.155$$

$$P=2^d.525 \qquad r_1(back)=0.163 \qquad r_2(back)=0.156$$

$$i=87^\circ.40 \qquad \Omega_1=7.093 \qquad V_\gamma=-8.7 \ km \ sec^{-1}$$

$$T_1=8500 \ K \qquad \Omega_2=6.926 \qquad F_1=1.00$$

$$T_2=8180 \ K \qquad q=0.91 \qquad F_2=1.00$$

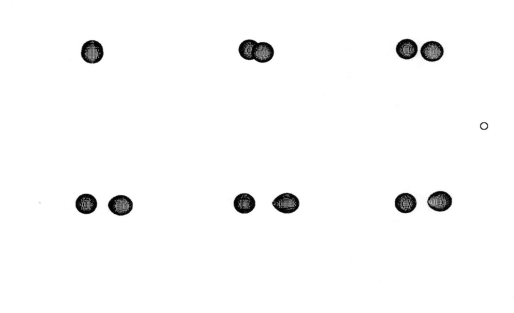

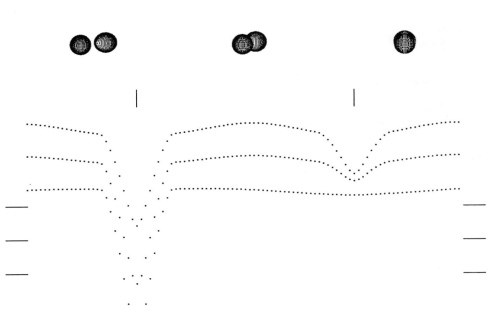

a=11.80 R$_\odot$ r$_1$(pole)=0.244   r$_2$(pole)=0.245

e= 0.000   r$_1$(point)=0.248   r$_2$(point)=0.358

ω= ———   r$_1$(side)=0.246   r$_2$(side)=0.255

P=3^d.3055   r$_1$(back)=0.247   r$_2$(back)=0.288

i=86°.0 Ω_1=4.336   V$_\gamma$=0.0 km sec^{-1}

T$_1$=7220 K Ω_2=2.330   F$_1$=1.00

T$_2$=4520 K   q=0.240   F$_2$=1.00

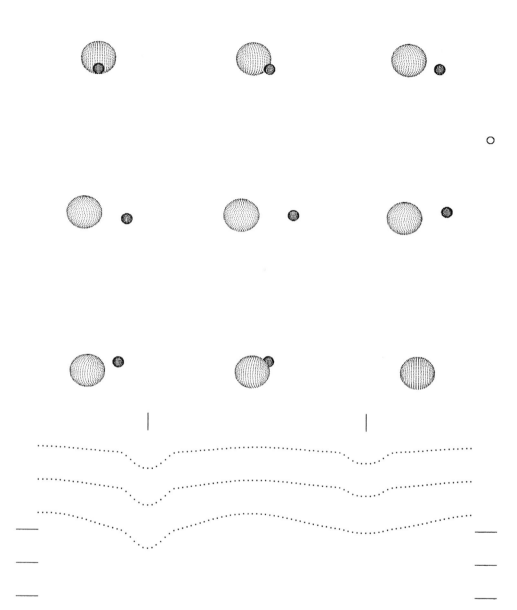

a=14.5 R$_\odot$	r$_1$(pole)=0.323	r$_2$(pole)=0.106
e= 0.000	r$_1$(point)=0.344	r$_2$(point)=0.106
ω= ---	r$_1$(side)=0.331	r$_2$(side)=0.106
P=2^d.9333	r$_1$(back)=0.339	r$_2$(back)=0.106
i=77°.6	Ω_1=3.548	V$_\gamma$=−4.0 km sec^{-1}
T$_1$=9630 K	Ω_2=5.701	F$_1$=1.00
T$_2$=7230 K	q=0.47	F$_2$=1.00

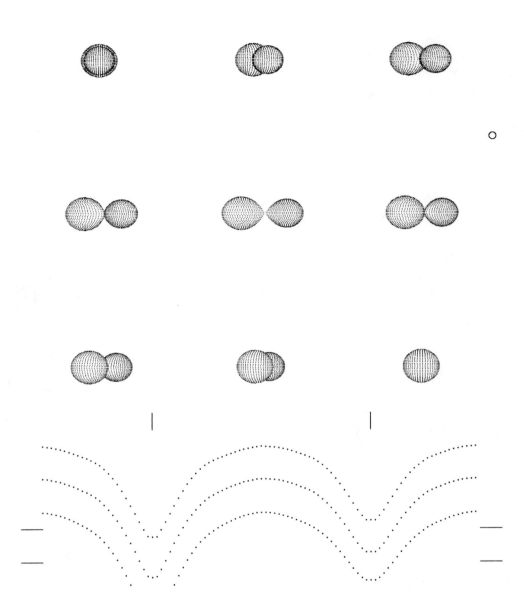

a=12.5 R$_\odot$ r$_1$(pole)=0.382   r$_2$(pole)=0.329

e= 0.000   r$_1$(point)=0.509   r$_2$(point)=0.467

ω= ---   r$_1$(side)=0.403   r$_2$(side)=0.345

P=1^d.1289   r$_1$(back)=0.432   r$_2$(back)=0.376

i=90°.0 Ω_1=3.291   V$_\gamma$= ---

T$_1$=25300 K Ω_2=3.282   F$_1$=1.00

T$_2$=20230 K   q=0.722   F$_2$=1.00

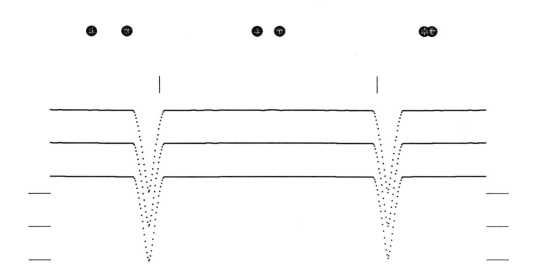

a= 13.5 R$_\odot$ r$_1$(pole)=0.110   r$_2$(pole)=0.112

e= 0.082   r$_1$(point)=0.110   r$_2$(point)=0.112

ω= 342°.3   r$_1$(side)=0.110   r$_2$(side)=0.112

P=3^d.8966   r$_1$(back)=0.110   r$_2$(back)=0.112

i=89°.9 Ω_1=10.182   V$_\gamma$= ———

T$_1$=6440 K Ω_2=10.028   F$_1$=1.18

T$_2$=6410 K   q=1.0   F$_2$=1.18

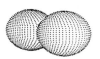

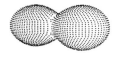 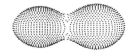 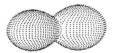

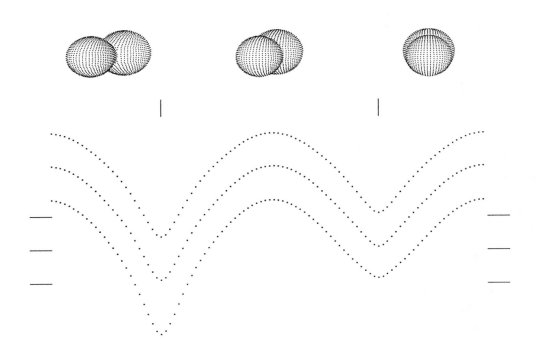

a=15.20 $R_\odot$ r_1(pole)=0.391 r_2(pole)=0.431

e= 0.000 r_1(point)=−1.000 r_2(point)=−1.000

ω= −−− r_1(side)=0.422 r_2(side)=0.468

P=1^d.6585 r_1(back)=0.514 r_2(back)=0.540

i=$85°$.05 Ω_1=3.751 V_γ=−27.7 km sec^{-1}

T_1=23000 K Ω_2=3.751 F_1=1.00

T_2=16182 K   q=1.28 F_2=1.00

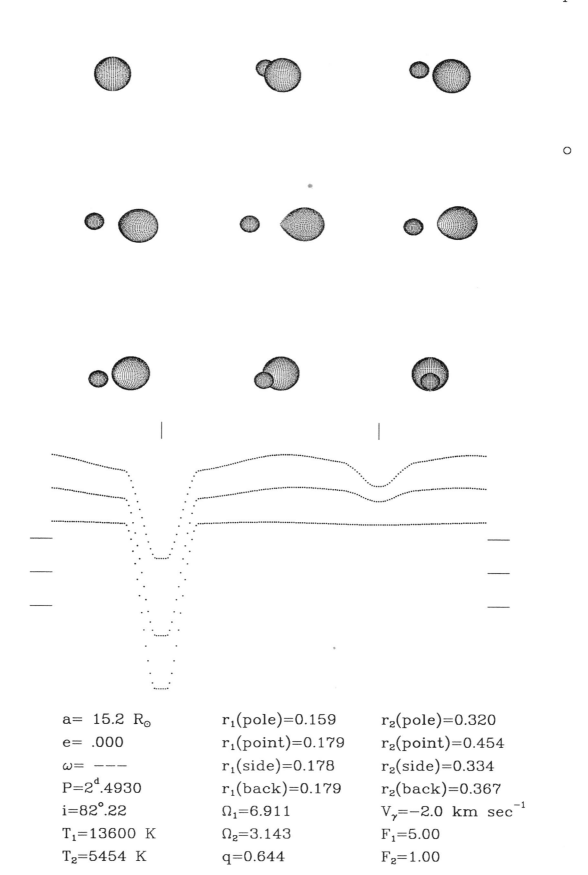

a= 15.2 $R_\odot$ r_1(pole)=0.159 r_2(pole)=0.320

e= .000 r_1(point)=0.179 r_2(point)=0.454

ω= ——— r_1(side)=0.178 r_2(side)=0.334

P=2^d.4930 r_1(back)=0.179 r_2(back)=0.367

i=$82°$.22 Ω_1=6.911 V_γ=−2.0 km sec^{-1}

T_1=13600 K Ω_2=3.143 F_1=5.00

T_2=5454 K   q=0.644 F_2=1.00

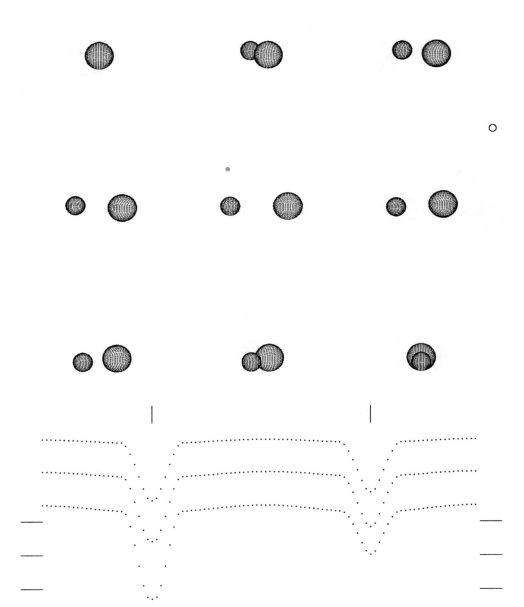

$a=$ 16.03 $R_\odot$ $r_1(\text{pole})=0.168$ $r_2(\text{pole})=0.244$

$e=$ 0.000 $r_1(\text{point})=0.170$ $r_2(\text{point})=0.255$

$\omega=$ ——— $r_1(\text{side})=0.169$ $r_2(\text{side})=0.248$

$P=3^{\text{d}}.3785$ $r_1(\text{back})=0.170$ $r_2(\text{back})=0.253$

$i=85°.4$ $\Omega_1=7.021$ $V_\gamma=-7$ km sec^{-1}

$T_1=9000$ K $\Omega_2=5.385$ $F_1=1.0$

$T_2=8170$ K $q=1.09$ $F_2=1.0$

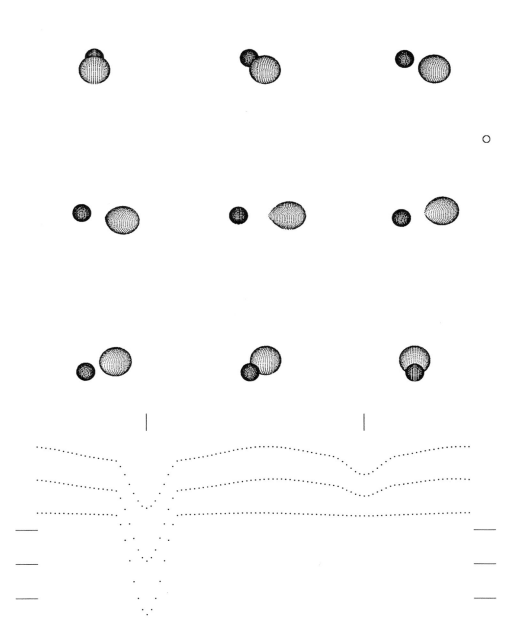

a= 13.99 $R_\odot$ r_1(pole)=0.183 r_2(pole)=0.295

e= 0.000 r_1(point)=0.185 r_2(point)=0.423

ω= ——— r_1(side)=0.184 r_2(side)=0.308

P=2^d.7746 r_1(back)=0.185 r_2(back)=0.340

i=75°.3 Ω_1=5.922 V_γ=−17.1 km sec^{-1}

T_1=13400 K Ω_2=2.812 F_1=1.0

T_2=6450 K   q=0.47 F_2=1.0

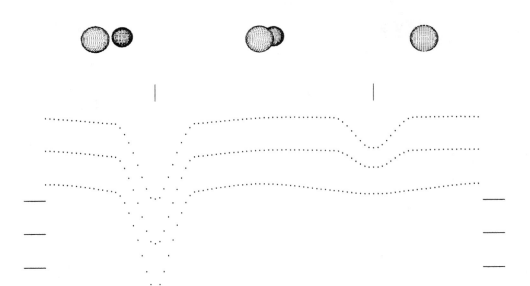

a= 12.71 $R_\odot$ r_1(pole)=0.302 r_2(pole)=0.224

e= 0.000 r_1(point)=0.317 r_2(point)=0.241

ω= --- r_1(side)=0.308 r_2(side)=0.228

P=2^d.1418 r_1(back)=0.314 r_2(back)=0.237

i=86°.3 Ω_1=3.753 V_γ=−17.4 km sec^{-1}

T_1=13510 K Ω_2=3.298 F_1=1.00

T_2=6480 K   q=0.46 F_2=1.00

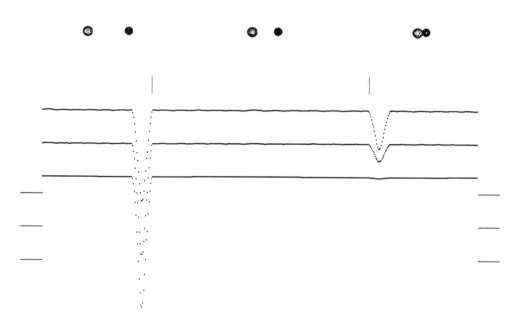

a= 16.61 R$_\odot$ r$_1$(pole)=0.086   r$_2$(pole)=0.071

e= 0.083   r$_1$(point)=0.086   r$_2$(point)=0.071

ω= 30°.3   r$_1$(side)=0.086   r$_2$(side)=0.071

P=4^d.4278   r$_1$(back)=0.086   r$_2$(back)=0.071

i=89°.4 Ω_1=12.233   V$_\gamma$=−10.9 km sec^{-1}

T$_1$=9610 K Ω_2=9.063   F$_1$=1.18

T$_2$=5830 K   q=0.55   F$_2$=1.18

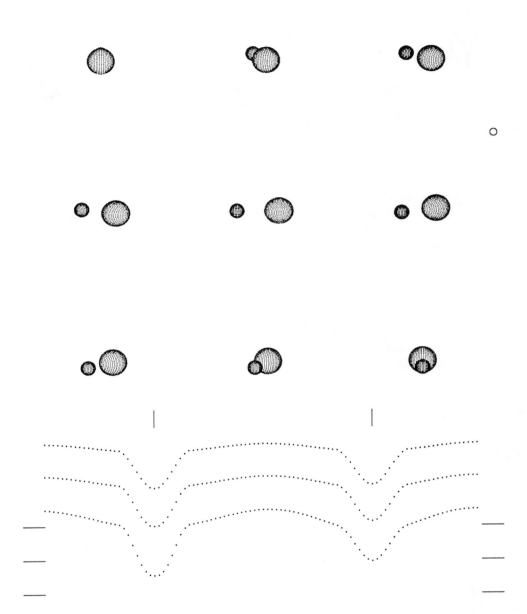

a= 11.7 R$_\odot$ r$_1$(pole)=0.174   r$_2$(pole)=0.320

e= 0.000   r$_1$(point)=0.178   r$_2$(point)=0.350

ω= ---   r$_1$(side)=0.175   r$_2$(side)=0.329

P=2^d.0583   r$_1$(back)=0.177   r$_2$(back)=0.341

i=80°.3 Ω_1=7.114   V$_\gamma$= ---

T$_1$=8500 K Ω_2=5.107   F$_1$=1.00

T$_2$=8262 K   q=1.39   F$_2$=1.00

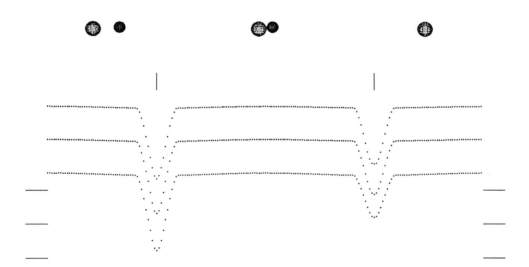

a=12.49 R$_\odot$ r$_1$(pole)=0.171   r$_2$(pole)=0.129

e= 0.000   r$_1$(point)=0.172   r$_2$(point)=0.130

ω= ———   r$_1$(side)=0.171   r$_2$(side)=0.129

P=2^d.7807   r$_1$(back)=0.172   r$_2$(back)=0.129

i=87°.5 Ω_1=6.795   V$_\gamma$=−7.9 km sec^{-1}

T$_1$=8700 K Ω_2=8.266   F$_1$=1.00

T$_2$=8100 K   q=0.932   F$_2$=1.00

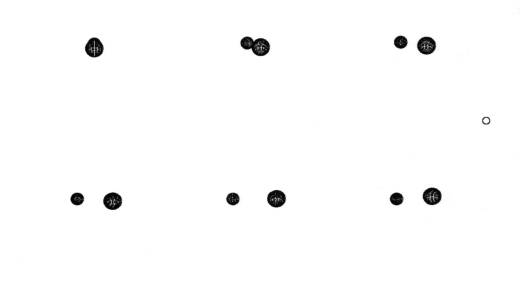

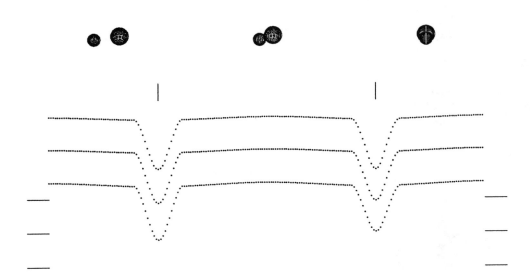

a=12.14 R$_\odot$ r$_1$(pole)=0.147   r$_2$(pole)=0.204

e= 0.000   r$_1$(point)=0.148   r$_2$(point)=0.209

ω= ———   r$_1$(side)=0.147   r$_2$(side)=0.206

P=2^d.8321   r$_1$(back)=0.148   r$_2$(back)=0.209

i=84°.0 Ω_1=7.820   V$_\gamma$=+20.0 km sec^{-1}

T$_1$=6410 K Ω_2=5.903   F$_1$=1.00

T$_2$=6360 K   q=1.006   F$_2$=1.00

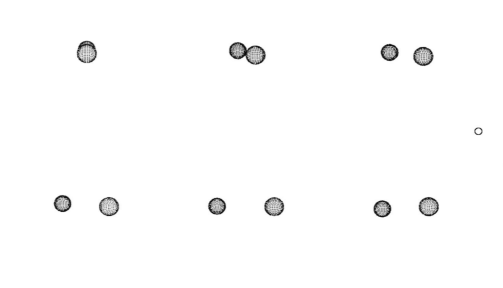

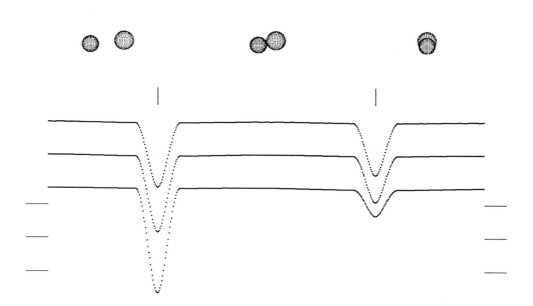

a= 17.65 R$_\odot$ r$_1$(pole)=0.150   r$_2$(pole)=0.159

e= 0.000   r$_1$(point)=0.151   r$_2$(point)=0.171

ω= ———   r$_1$(side)=0.150   r$_2$(side)=0.170

P=5^d.1171   r$_1$(back)=0.151   r$_2$(back)=0.171

i=84°.6 Ω_1=7.688   V$_\gamma$=−4 km sec^{-1}

T$_1$=5800 K Ω_2=6.951   F$_1$=1.00

T$_2$=5200 K   q=1.01   F$_2$=1.00

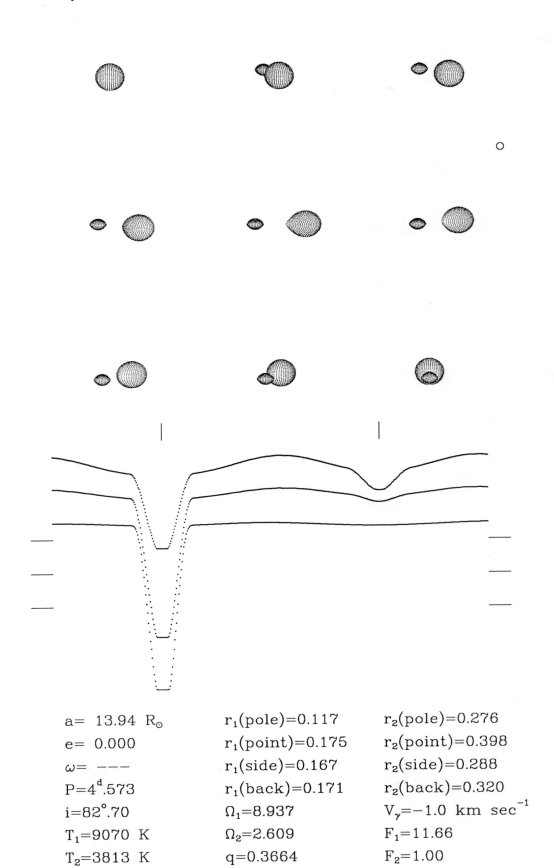

a= 13.94 R$_\odot$	r$_1$(pole)=0.117	r$_2$(pole)=0.276
e= 0.000	r$_1$(point)=0.175	r$_2$(point)=0.398
ω= ---	r$_1$(side)=0.167	r$_2$(side)=0.288
P=4^d.573	r$_1$(back)=0.171	r$_2$(back)=0.320
i=82°.70	Ω_1=8.937	V$_\gamma$=−1.0 km sec^{-1}
T$_1$=9070 K	Ω_2=2.609	F$_1$=11.66
T$_2$=3813 K	q=0.3664	F$_2$=1.00

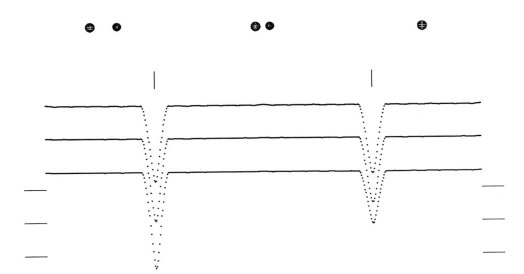

a= 12.69 R$_\odot$	r$_1$(pole)=0.104	r$_2$(pole)=0.090
e= 0.000	r$_1$(point)=0.104	r$_2$(point)=0.090
ω= ---	r$_1$(side)=0.104	r$_2$(side)=0.090
P=3^d.2613	r$_1$(back)=0.104	r$_2$(back)=0.090
i=89°.32	Ω_1=10.510	V$_\gamma$=−16.2 km sec^{-1}
T$_1$=6100 K	Ω_2=11.223	F$_1$=1.00
T$_2$=5844 K	q=0.92	F$_2$=1.00

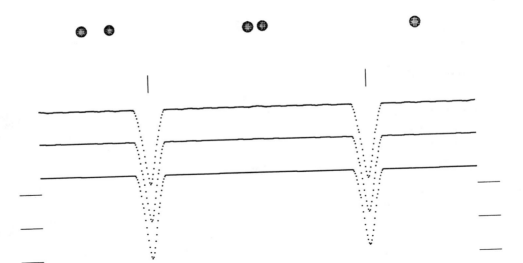

a= 13.18 R$_\odot$ r$_1$(pole)=0.114   r$_2$(pole)=0.107

e= 0.000   r$_1$(point)=0.114   r$_2$(point)=0.108

ω= ———   r$_1$(side)=0.114   r$_2$(side)=0.108

P=3^d.3640   r$_1$(back)=0.114   r$_2$(back)=0.108

i=89°.02 Ω_1=9.758   V$_\gamma$= +1.3 km sec^{-1}

T$_1$=6750 K Ω_2=10.223   F$_1$=1.0

T$_2$=6709 K   q=0.999   F$_2$=1.0

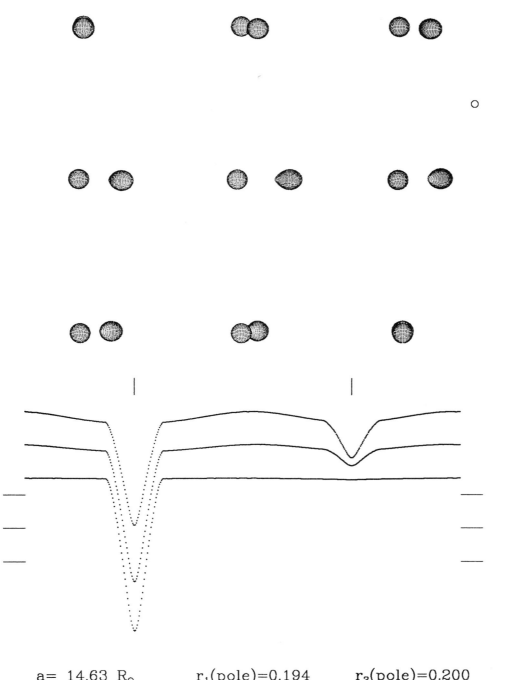

a= 14.63 $R_\odot$ r_1(pole)=0.194 r_2(pole)=0.200

e= 0.000 r_1(point)=0.195 r_2(point)=0.297

ω= ——— r_1(side)=0.195 r_2(side)=0.208

P=3^d.4361 r_1(back)=0.195 r_2(back)=0.240

i=87°.4 Ω_1=5.266 V_γ=−48.0 km sec^{-1}

T_1=11400 K Ω_2=2.019 F_1=1.00

T_2=5690 K   q=0.12 F_2=1.00

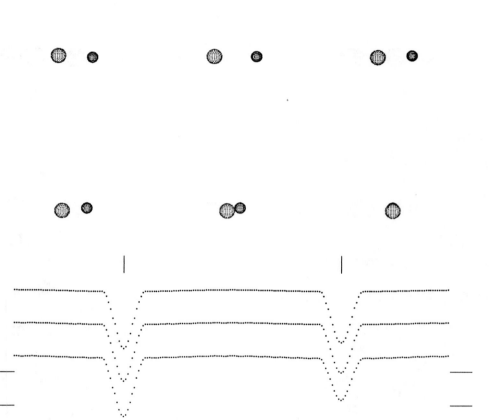

a=11.72 R$_\odot$ r$_1$(pole)=0.181   r$_2$(pole)=0.132

e= 0.000   r$_1$(point)=0.184   r$_2$(point)=0.133

ω= ---   r$_1$(side)=0.182   r$_2$(side)=0.132

P=2^d.7284   r$_1$(back)=0.183   r$_2$(back)=0.132

i=85°.7 Ω_1=6.351   V$_\gamma$=+36.1 km sec^{-1}

T$_1$=7200 K Ω_2=7.451   F$_1$=1.00

T$_2$=6920 K   q=0.84   F$_2$=1.00

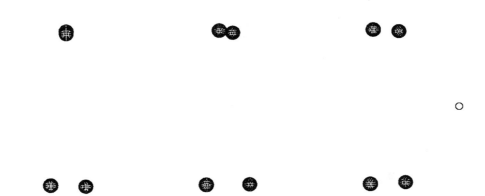

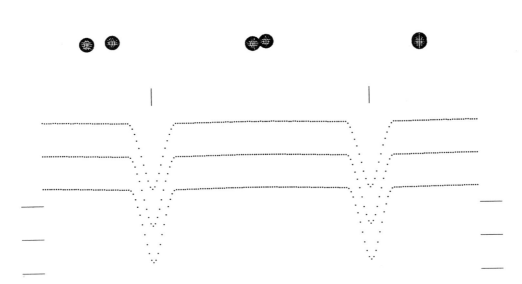

a= 12.07 R$_\odot$	r$_1$(pole)=0.170	r$_2$(pole)=0.165
e= 0.000	r$_1$(point)=0.172	r$_2$(point)=0.167
ω= ---	r$_1$(side)=0.171	r$_2$(side)=0.166
P=2^d.9685	r$_1$(back)=0.172	r$_2$(back)=0.167
i=86°.81	Ω_1=6.853	V$_\gamma$= +16.2 km sec^{-1}
T$_1$=6500 K	Ω_2=6.981	F$_1$=1.00
T$_2$=6500 K	q=0.99	F$_2$=1.00

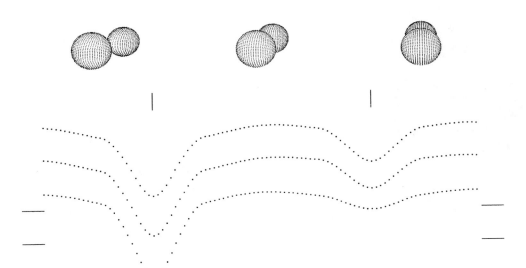

a= 14.90 $R_\odot$	r_1(pole)=0.359	r_2(pole)=0.281
e= 0.000	r_1(point)=0.391	r_2(point)=0.404
ω= ---	r_1(side)=0.372	r_2(side)=0.292
P=2^d.0510	r_1(back)=0.382	r_2(back)=0.325
i=78°.0	Ω_1=3.150	V_γ=−11.8 km sec^{-1}
T_1=20000 K	Ω_2=2.658	F_1=1.00
T_2=11000 K	q=0.39	F_2=1.00

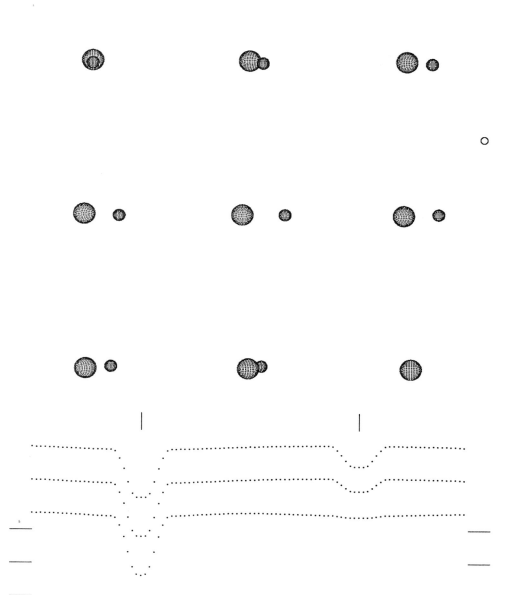

a= 11.92 $R_{\odot}$ r_1(pole)=0.251 r_2(pole)=0.143

e= 0.000 r_1(point)=0.256 r_2(point)=0.146

ω= --- r_1(side)=0.254 r_2(side)=0.143

P=1^d.6374 r_1(back)=0.256 r_2(back)=0.146

i=$86°$.6 Ω_1=4.259 V_{γ}= −16.0 km sec^{-1}

T_1=15000 K Ω_2=3.377 F_1=1.0

T_2=7510 K   q=0.29 F_2=1.0

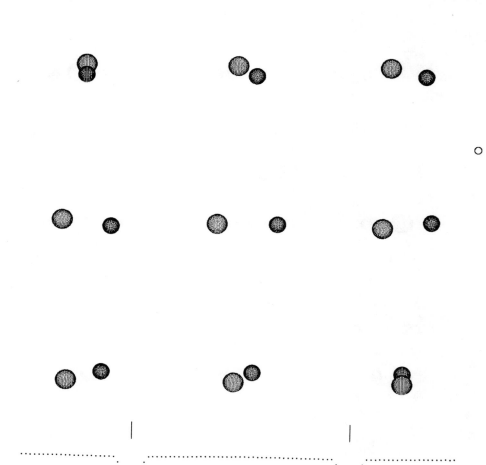

a= 16.73 R$_\odot$	r$_1$(pole)=0.169	r$_2$(pole)=0.139
e= 0.000	r$_1$(point)=0.171	r$_2$(point)=0.140
ω= ---	r$_1$(side)=0.170	r$_2$(side)=0.139
P=3^d.8950	r$_1$(back)=0.171	r$_2$(back)=0.140
i=79°.5	Ω_1=6.720	V$_\gamma$=−38.6km sec^{-1}
T$_1$=8790 K	Ω_2=7.024	F$_1$=1.00
T$_2$=8600 K	q=0.824	F$_2$=1.00

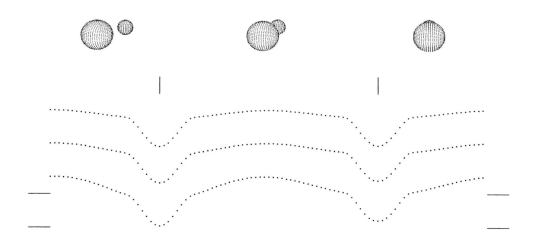

a=13.36 $R_\odot$ r_1(pole)=0.322 r_2(pole)=0.161

e=0.000 r_1(point)=0.356 r_2(point)=0.163

ω= --- r_1(side)=0.332 r_2(side)=0.161

P=2^d.2677 r_1(back)=0.345 r_2(back)=0.163

i=78°.4 Ω_1=3.802 V_γ=+29.3 km sec^{-1}

T_1=11750 K Ω_2=5.675 F_1=1.00

T_2=11100 K   q=0.73 F_2=1.00

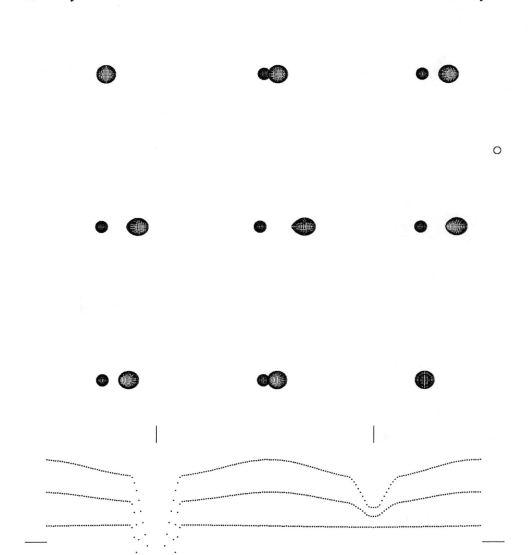

$$a=12.4\ R_\odot \qquad r_1(\text{pole})=0.139 \qquad r_2(\text{pole})=0.209$$

$$e=\ 0.000 \qquad r_1(\text{point})=0.139 \qquad r_2(\text{point})=0.309$$

$$\omega=\ --- \qquad r_1(\text{side})=0.139 \qquad r_2(\text{side})=0.217$$

$$P=3^d.5359 \qquad r_1(\text{back})=0.139 \qquad r_2(\text{back})=0.249$$

$$i=89°.5 \qquad \Omega_1=7.342 \qquad V_\gamma=\ ---$$

$$T_1=8620\ K \qquad \Omega_2=2.070 \qquad F_1=1.00$$

$$T_2=4700\ K \qquad q=0.138 \qquad F_2=1.00$$

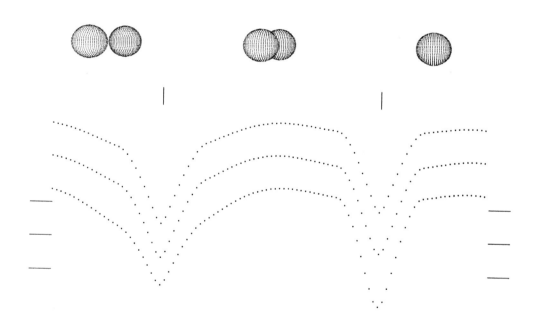

a= 16.70 $R_\odot$ r_1(pole)=0.284 r_2(pole)=0.266

e= 0.000 r_1(point)=0.353 r_2(point)=0.275

ω= ——— r_1(side)=0.295 r_2(side)=0.270

P=5^d.0740 r_1(back)=0.322 r_2(back)=0.273

i=89°.0 Ω_1=5.579 V_γ= −53.6 km sec^{-1}

T_1=5300 K Ω_2=8.436 F_1=1.00

T_2=5715 K   q=2.14 F_2=1.00

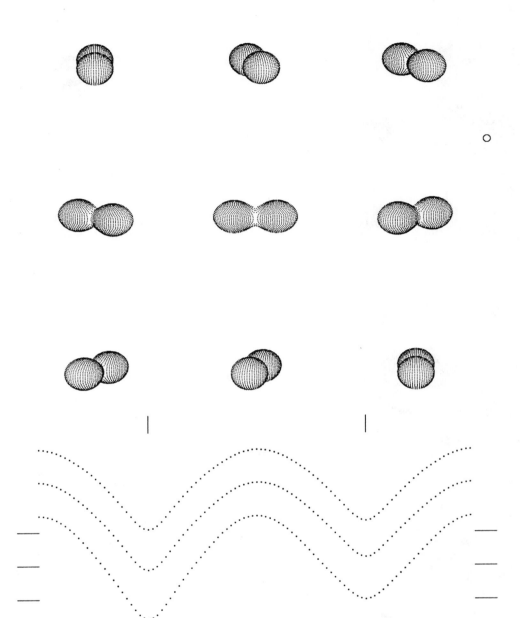

a= 11.5 R$_\odot$ r$_1$(pole)=0.410   r$_2$(pole)=0.410

e= 0.000   r$_1$(point)=−1.000   r$_2$(point)=−1.000

ω= −−−   r$_1$(side)=0.444   r$_2$(side)=0.444

P=1^d.1428   r$_1$(back)=0.522   r$_2$(back)=0.522

i=78°.45 Ω_1=3.365   V$_\gamma$= −−−

T$_1$=20500 K Ω_2=3.365   F$_1$=1.0

T$_2$=16754 K   q=1.0   F$_2$=1.0

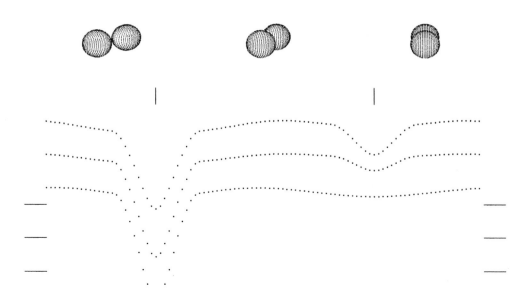

$a = 13.72\ R_\odot$ $r_1(\text{pole}) = 0.292$ $r_2(\text{pole}) = 0.272$

$e = 0.000$ $r_1(\text{point}) = 0.303$ $r_2(\text{point}) = 0.394$

$\omega = \ ---$ $r_1(\text{side}) = 0.297$ $r_2(\text{side}) = 0.284$

$P = 2^d.3274$ $r_1(\text{back}) = 0.301$ $r_2(\text{back}) = 0.317$

$i = 80°.9$ $\Omega_1 = 3.762$ $V_\gamma = -40.0\ \text{km sec}^{-1}$

$T_1 = 10160\ K$ $\Omega_2 = 2.574$ $F_1 = 1.00$

$T_2 = 5350\ K$ $q = 0.35$ $F_2 = 1.00$

$$a=14.67\ R_\odot \qquad r_1(pole)=0.092 \qquad r_2(pole)=0.087$$

$$e=0.003 \qquad r_1(point)=0.092 \qquad r_2(point)=0.087$$

$$\omega=71°.63 \qquad r_1(side)=0.092 \qquad r_2(side)=0.087$$

$$P=4^d.1811 \qquad r_1(back)=0.092 \qquad r_2(back)=0.087$$

$$i=89°.6 \qquad \Omega_1=11.869 \qquad V_\gamma=+48.5\ km\ sec^{-1}$$

$$T_1=6195\ K \qquad \Omega_2=12.166 \qquad F_1=1.01$$

$$T_2=6152\ K \qquad q=0.968 \qquad F_2=1.01$$

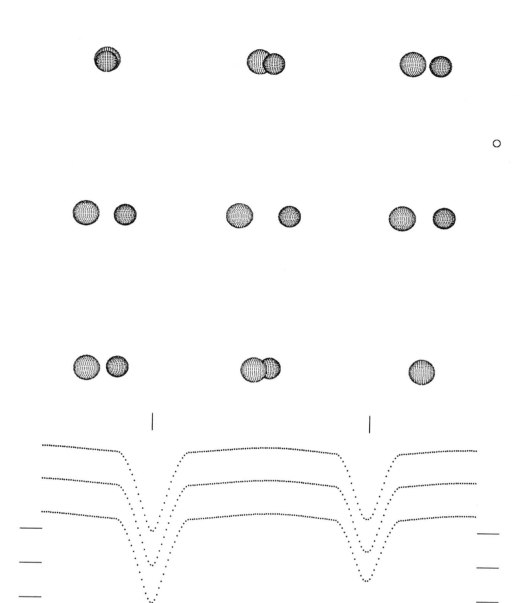

a=14.14 R$_\odot$ r$_1$(pole)=0.252   r$_2$(pole)=0.215

e=0.059   r$_1$(point)=0.268   r$_2$(point)=0.224

ω=261°.18   r$_1$(side)=0.257   r$_2$(side)=0.218

P=1^d.8848   r$_1$(back)=0.264   r$_2$(back)=0.222

i=87°.8 Ω_1=4.946   V$_\gamma$=+15.0 km sec^{-1}

T$_1$=18730 K Ω_2=5.476   F$_1$=1.13

T$_2$=17150 K   q=0.95   F$_2$=1.13

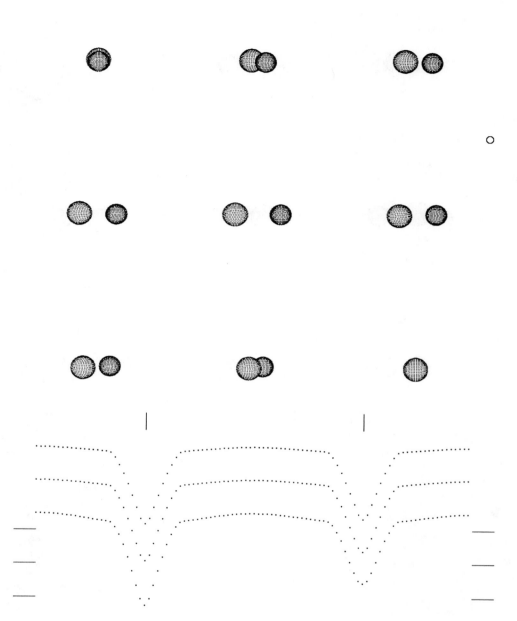

a= 12.65 $R_\odot$ r_1(pole)=0.267 r_2(pole)=0.232

e= 0.000 r_1(point)=0.283 r_2(point)=0.242

ω= ——— r_1(side)=0.272 r_2(side)=0.235

P=1^d.6773 r_1(back)=0.279 r_2(back)=0.240

i=87°.7 Ω_1=4.635 V_γ=−12 km sec^{-1}

T_1=16000 K Ω_2=4.981 F_1=1.00

T_2=15070 K   q=0.92 F_2=1.00

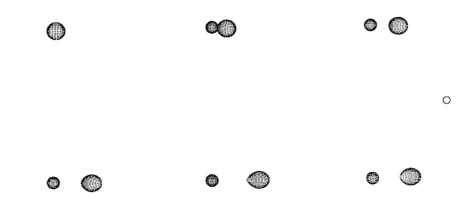

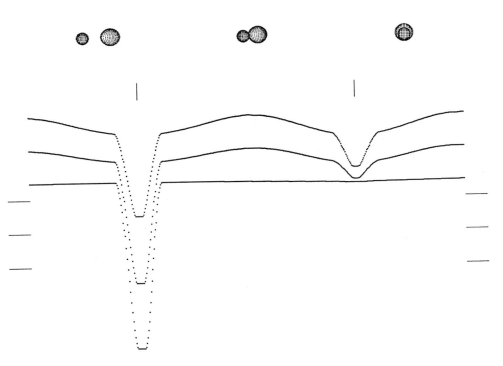

$$a= 13.1 \ R_\odot \qquad r_1(pole)=0.134 \qquad r_2(pole)=0.190$$

$$e= 0.000 \qquad r_1(point)=0.134 \qquad r_2(point)=0.282$$

$$\omega= \ --- \qquad r_1(side)=0.134 \qquad r_2(side)=0.197$$

$$P=3^d.6871 \qquad r_1(back)=0.134 \qquad r_2(back)=0.228$$

$$i=88°.1 \qquad \Omega_1=7.571 \qquad V_\gamma= \ ---$$

$$T_1=9400 \ K \qquad \Omega_2=1.959 \qquad F_1=1.00$$

$$T_2=5370 \ K \qquad q=0.10 \qquad F_2=1.00$$

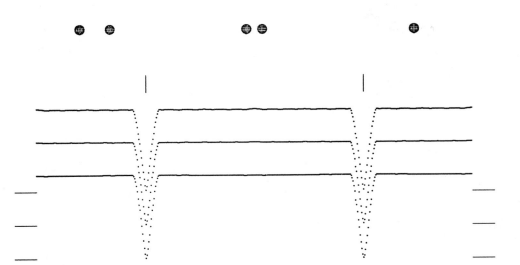

a= 14.28 $R_\odot$ r_1(pole)=0.096 r_2(pole)=0.093

e= 0.000 r_1(point)=0.096 r_2(point)=0.093

ω= --- r_1(side)=0.096 r_2(side)=0.093

P=4^d.1835 r_1(back)=0.096 r_2(back)=0.093

i=89°.0 Ω_1=11.391 V_γ= −24 km sec^{-1}

T_1=6200 K Ω_2=11.632 F_1=1.00

T_2=6220 K   q=0.99 F_2=1.00

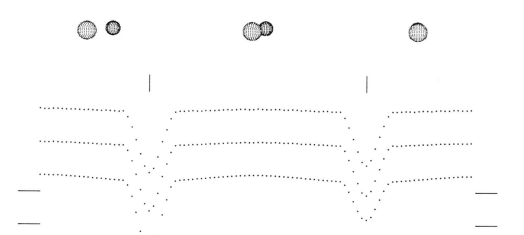

a= 12.26 R$_\odot$ r$_1$(pole)=0.213   r$_2$(pole)=0.165

e= 0.0125   r$_1$(point)=0.219   r$_2$(point)=0.168

ω= 254°.0   r$_1$(side)=0.215   r$_2$(side)=0.166

P=2^d.1967   r$_1$(back)=0.218   r$_2$(back)=0.167

i=85°.90 Ω_1=5.535   V$_\gamma$= −9.9 km sec^{-1}

T$_1$=10800 K Ω_2=6.219   F$_1$=1.02

T$_2$=9800 K   q=0.85   F$_2$=1.02

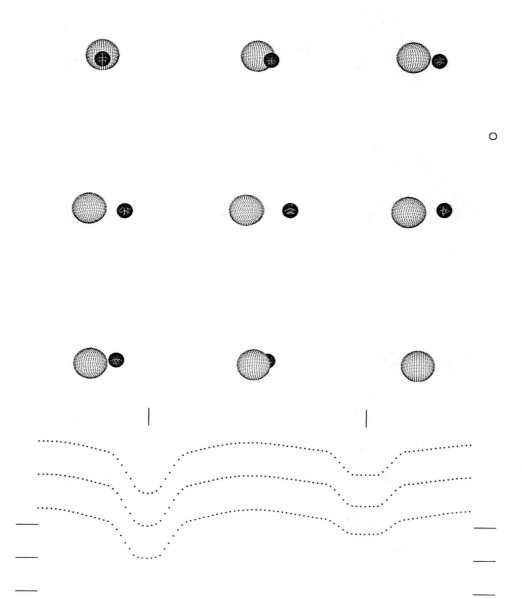

$$a=12.21 \ R_\odot \qquad r_1(pole)=0.365 \qquad r_2(pole)=0.173$$

$$e= 0.000 \qquad r_1(point)=0.405 \qquad r_2(point)=0.178$$

$$\omega= \ \text{---} \qquad r_1(side)=0.379 \qquad r_2(side)=0.175$$

$$P=1^d.4854 \qquad r_1(back)=0.392 \qquad r_2(back)=0.177$$

$$i=84°.5 \qquad \Omega_1=3.172 \qquad V_\gamma=+22.1 \ km \ sec^{-1}$$

$$T_1=21000 \ K \qquad \Omega_2=3.903 \qquad F_1=1.00$$

$$T_2=15200 \ K \qquad q=0.458 \qquad F_2=1.00$$

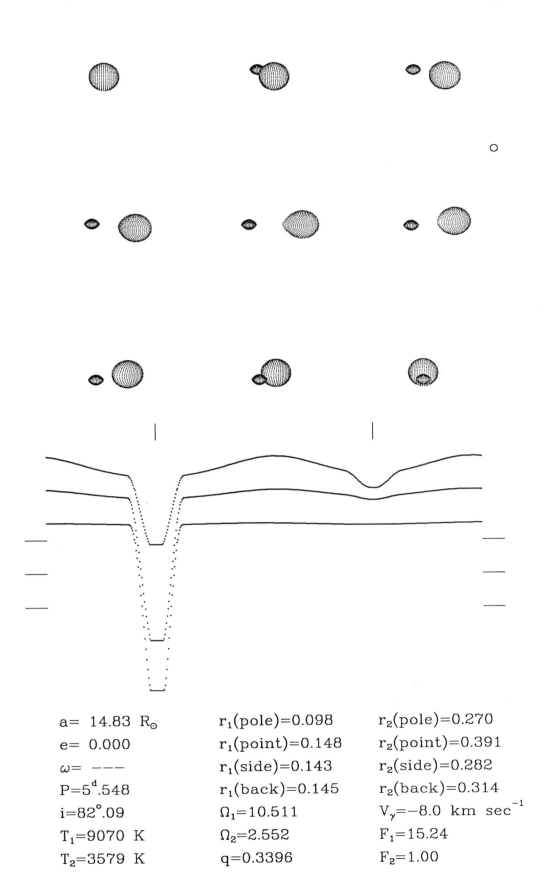

a= 14.83 $R_\odot$	r_1(pole)=0.098	r_2(pole)=0.270
e= 0.000	r_1(point)=0.148	r_2(point)=0.391
ω= ---	r_1(side)=0.143	r_2(side)=0.282
P=5^d.548	r_1(back)=0.145	r_2(back)=0.314
i=82°.09	Ω_1=10.511	V_γ=−8.0 km sec^{-1}
T_1=9070 K	Ω_2=2.552	F_1=15.24
T_2=3579 K	q=0.3396	F_2=1.00

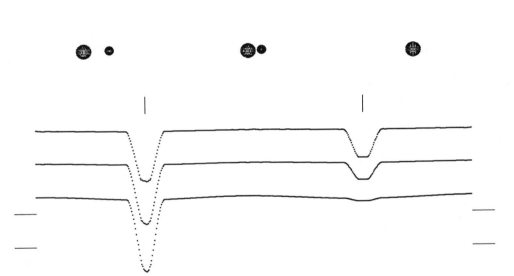

a= 12.14 R$_\odot$ r$_1$(pole)=0.171   r$_2$(pole)=0.102

e= 0.000   r$_1$(point)=0.173   r$_2$(point)=0.102

ω= ———   r$_1$(side)=0.172   r$_2$(side)=0.102

P=2^d.6282   r$_1$(back)=0.173   r$_2$(back)=0.102

i=88°.4 Ω_1=6.448   V$_\gamma$= variable

T$_1$=8510 K Ω_2=7.272   F$_1$=1.00

T$_2$=6170 K   q=0.62   F$_2$=1.00

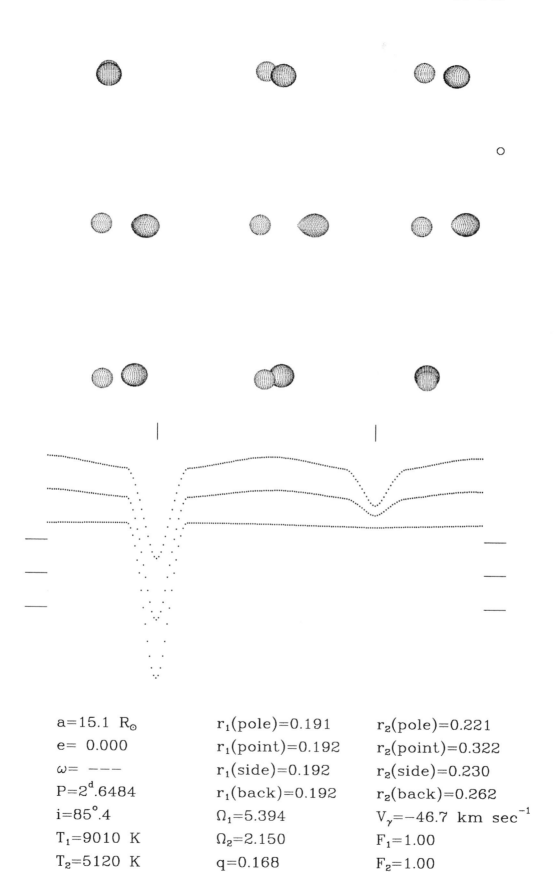

a=15.1 $R_\odot$ $r_1(pole)=0.191$ $r_2(pole)=0.221$

e= 0.000 $r_1(point)=0.192$ $r_2(point)=0.322$

$\omega=$ --- $r_1(side)=0.192$ $r_2(side)=0.230$

$P=2^d.6484$ $r_1(back)=0.192$ $r_2(back)=0.262$

$i=85^\circ.4$ $\Omega_1=5.394$ $V_\gamma=-46.7$ km sec^{-1}

$T_1=9010$ K $\Omega_2=2.150$ $F_1=1.00$

$T_2=5120$ K $q=0.168$ $F_2=1.00$

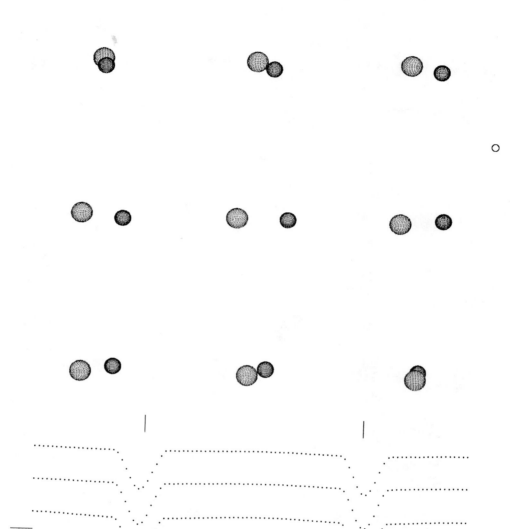

a= 13.837 $R_\odot$ r_1(pole)=0.211 r_2(pole)=0.164

e= 0.053 r_1(point)=0.218 r_2(point)=0.167

ω= 297°.07 r_1(side)=0.214 r_2(side)=0.165

P=2^d.0287 r_1(back)=0.217 r_2(back)=0.166

i=81°.36 Ω_1=5.670 V_γ=+25.8 km sec^{-1}

T_1=16930 K Ω_2=6.628 F_1=1.11

T_2=15683 K   q=0.91 F_2=1.11

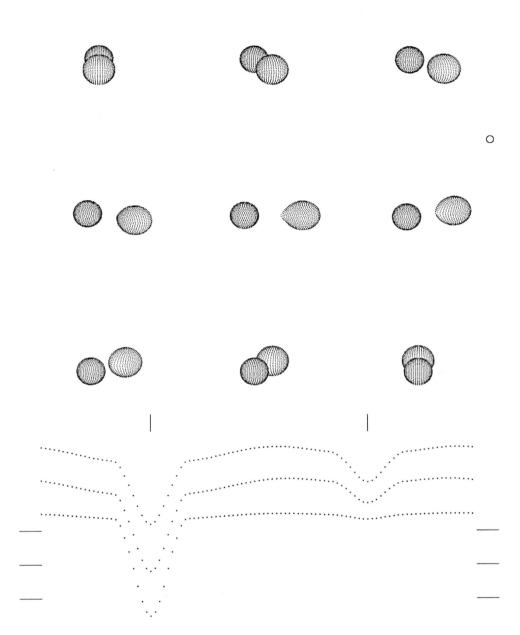

a=16.17 $R_\odot$ r_1(pole)=0.243 r_2(pole)=0.266

e= 0.000 r_1(point)=0.248 r_2(point)=0.385

ω= --- r_1(side)=0.245 r_2(side)=0.277

P=2^d.7277 r_1(back)=0.247 r_2(back)=0.310

i=78°.7 Ω_1=4.426 V_γ=−13.2 kms sec^{-1}

T_1=15900 K Ω_2=2.510 F_1=1.00

T_2=8224 K   q=0.32 F_2=1.00

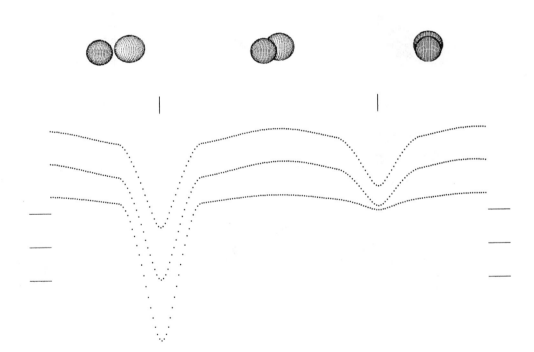

$a=14.1\ R_\odot$ $r_1(pole)=0.257$ $r_2(pole)=0.278$

$e=0.000$ $r_1(point)=0.262$ $r_2(point)=0.400$

$\omega=$ ——— $r_1(side)=0.260$ $r_2(side)=0.290$

$P=3^d.6877$ $r_1(back)=0.262$ $r_2(back)=0.323$

$i=85°.1$ $\Omega_1=4.259$ $V_\gamma=-51.2\ km\ sec^{-1}$

$T_1=12220\ K$ $\Omega_2=2.633$ $F_1=1.00$

$T_2=8130\ K$ $q=0.38$ $F_2=1.00$

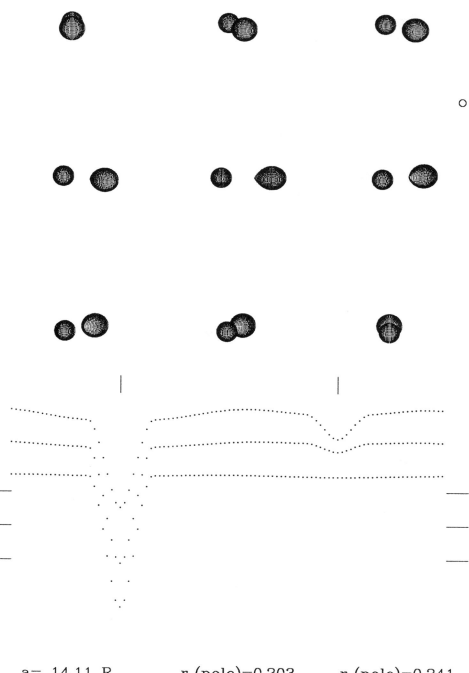

a= 14.11 R$_\odot$ r$_1$(pole)=0.203   r$_2$(pole)=0.241

e= 0.000   r$_1$(point)=0.205   r$_2$(point)=0.353

ω= ———   r$_1$(side)=0.204   r$_2$(side)=0.251

P=2^d.867   r$_1$(back)=0.205   r$_2$(back)=0.284

i=82°.31 Ω_1=5.151   V$_\gamma$= variable

T$_1$=12000 K Ω_2=2.299   F$_1$=1.00

T$_2$=4888 K   q=0.227   F$_2$=1.00

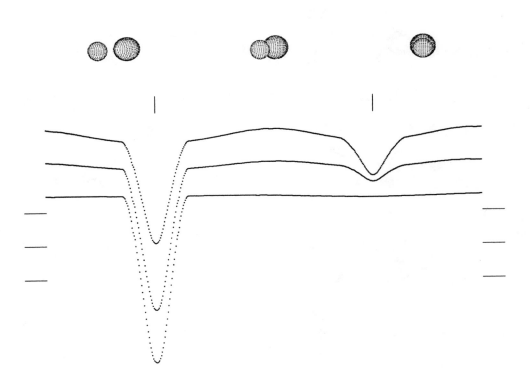

$a=\ 13.80\ R_{\odot}$ $r_1(\text{pole})=0.198$ $r_2(\text{pole})=0.245$

$e=\ 0.000$ $r_1(\text{point})=0.200$ $r_2(\text{point})=0.358$

$\omega=\ ---$ $r_1(\text{side})=0.199$ $r_2(\text{side})=0.255$

$P=3^{d}.7659$ $r_1(\text{back})=0.199$ $r_2(\text{back})=0.288$

$i=87°.1$ $\Omega_1=5.285$ $V_{\gamma}=\ +6\ \text{km sec}^{-1}$

$T_1=9300\ K$ $\Omega_2=2.329$ $F_1=1.00$

$T_2=4860\ K$ $q=0.24$ $F_2=1.00$

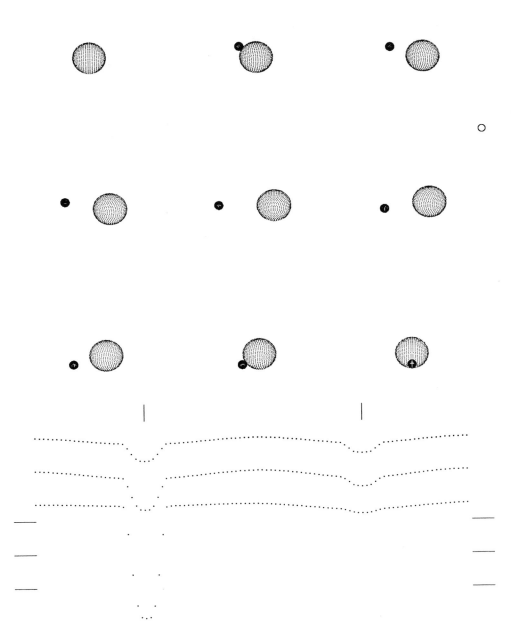

a= 15.38 $R_\odot$ r_1(pole)=0.080 r_2(pole)=0.293

e= 0.000 r_1(point)=0.080 r_2(point)=0.314

ω= ——— r_1(side)=0.080 r_2(side)=0.300

P=3^d.9653 r_1(back)=0.080 r_2(back)=0.308

i=77°.75 Ω_1=13.842 V_γ= +8.8 km sec^{-1}

T_1=6100 K Ω_2=5.360 F_1=1.00

T_2=4700 K   q=1.34 F_2=1.00

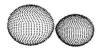 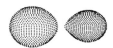 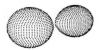

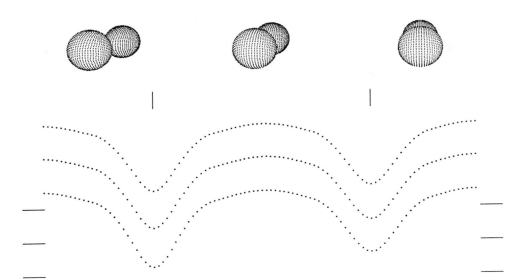

a=15.26 $R_\odot$ r_1(pole)=0.386 r_2(pole)=0.304

e= 0.000 r_1(point)=0.456 r_2(point)=0.435

ω= --- r_1(side)=0.405 r_2(side)=0.318

P=1^d.4545 r_1(back)=0.426 r_2(back)=0.350

i=79°.0 Ω_1=3.085 V_γ=+13.2 km sec^{-1}

T_1=27000 K Ω_2=2.933 F_1=1.00

T_2=24354 K   q=0.53 F_2=1.00

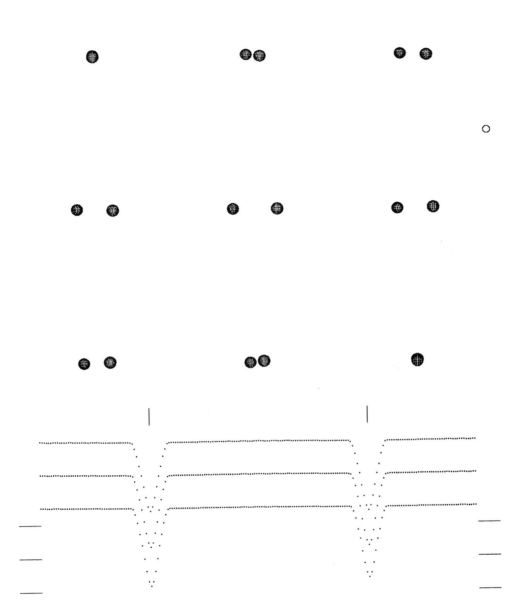

$a=12.25\ R_{\odot}$ $r_1(pole)=0.130$ $r_2(pole)=0.137$

$e=\ 0.000$ $r_1(point)=0.131$ $r_2(point)=0.138$

$\omega=\ ---$ $r_1(side)=0.130$ $r_2(side)=0.137$

$P=3^d.1986$ $r_1(back)=0.130$ $r_2(back)=0.138$

$i=87°.88$ $\Omega_1=8.678$ $V_\gamma=+63.2\ km\ sec^{-1}$

$T_1=5400\ K$ $\Omega_2=8.171$ $F_1=1.00$

$T_2=5340\ K$ $q=0.98$ $F_2=1.00$

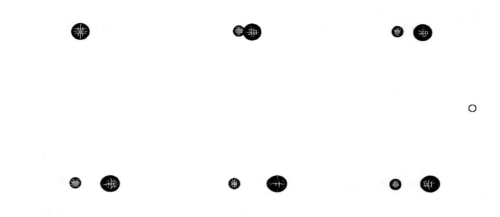

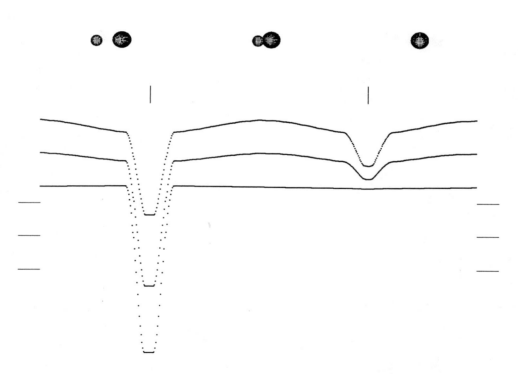

a= 11.9 R$_\odot$	r$_1$(pole)=0.132	r$_2$(pole)=0.199
e= 0.000	r$_1$(point)=0.132	r$_2$(point)=0.246
ω= ---	r$_1$(side)=0.132	r$_2$(side)=0.206
P=3^d.2756	r$_1$(back)=0.132	r$_2$(back)=0.229
i=88°.5	Ω_1=7.706	V$_\gamma$= +8.2 km sec^{-1}
T$_1$=9200 K	Ω_2=2.112	F$_1$=1.00
T$_2$=5150 K	q=0.14	F$_2$=1.00

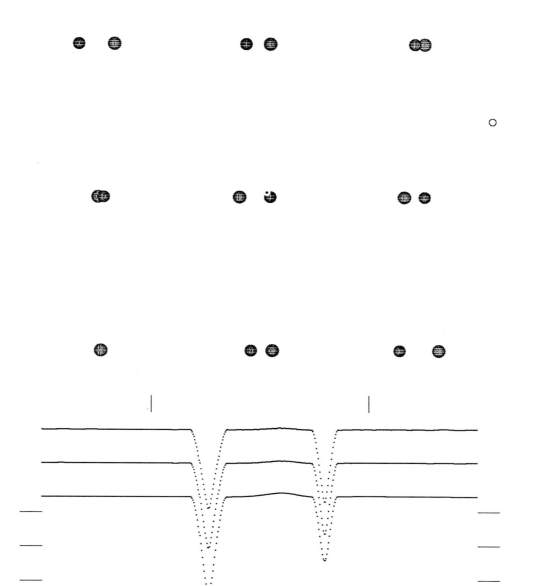

a= 14.96 R$_\odot$	r$_1$(pole)=0.120	r$_2$(pole)=0.110
e= 0.41	r$_1$(point)=0.124	r$_2$(point)=0.112
ω= 205°.20	r$_1$(side)=0.122	r$_2$(side)=0.111
P=3^d.2828	r$_1$(back)=0.123	r$_2$(back)=0.112
i=90°.0	Ω_1=9.793	V$_\gamma$=−16.8 km sec^{-1}
T$_1$=10000 K	Ω_2=9.904	F$_1$=2.62
T$_2$=9500 K	q=0.90	F$_2$=2.62

O

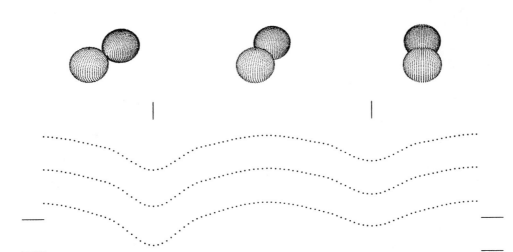

a= 14.89 $R_\odot$	r_1(pole)=0.345	r_2(pole)=0.324
e= 0.000	r_1(point)=0.389	r_2(point)=0.460
ω= ---	r_1(side)=0.358	r_2(side)=0.338
P=1^d.4463	r_1(back)=0.373	r_2(back)=0.370
i=63°.18	Ω_1=3.537	V_γ= 0 km sec^{-1}
T_1=21500 K	Ω_2=3.198	F_1=1.00
T_2=16242 K	q=0.674	F_2=1.00

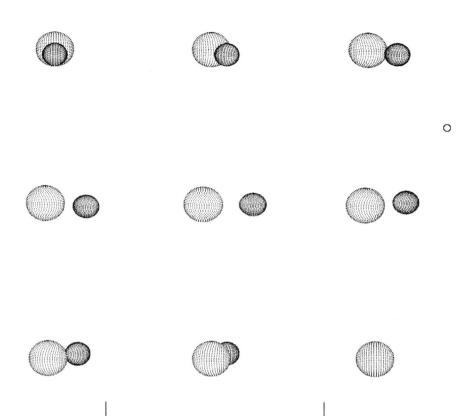

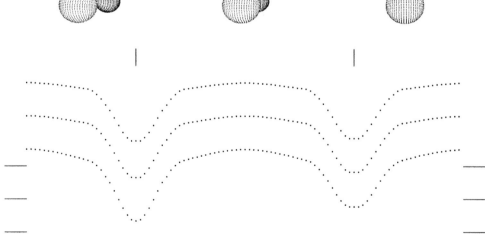

a= 13.9 $R_\odot$	r_1(pole)=0.367	r_2(pole)=0.240
e= 0.000	r_1(point)=0.397	r_2(point)=0.283
ω= ---	r_1(side)=0.380	r_2(side)=0.248
P=2^d.3333	r_1(back)=0.389	r_2(back)=0.268
i=83°.52	Ω_1=3.009	V_γ= ---
T_1=16400 K	Ω_2=2.593	F_1=1.00
T_2=15788 K	q=0.306	F_2=1.00

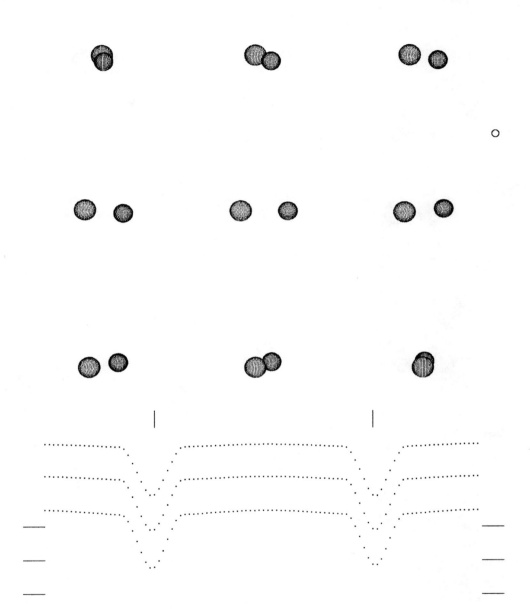

a= 12.88 R$_\odot$ r$_1$(pole)=0.231   r$_2$(pole)=0.204

e= 0.027   r$_1$(point)=0.240   r$_2$(point)=0.210

ω= 311°.0   r$_1$(side)=0.234   r$_2$(side)=0.205

P=1^d.7309   r$_1$(back)=0.238   r$_2$(back)=0.208

i=82°.17 Ω_1=5.600   V$_\gamma$=+0.5 km sec^{-1}

T$_1$=16900 K Ω_2=5.256   F$_1$=1.06

T$_2$=16340 K   q=0.928   F$_2$=1.06

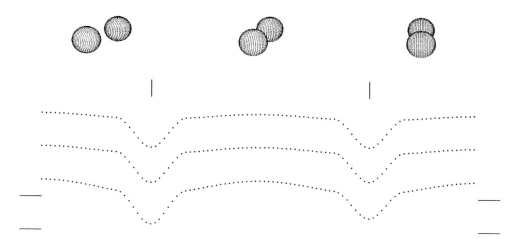

a= 14.76 $R_\odot$ r_1(pole)=0.267 r_2(pole)=0.246

e= 0.000 r_1(point)=0.283 r_2(point)=0.260

ω= ——— r_1(side)=0.272 r_2(side)=0.250

P=2^d.7858 r_1(back)=0.279 r_2(back)=0.257

i=73°.93 Ω_1=4.594 V_γ=−32 km sec^{-1}

T_1=10700 K Ω_2=4.618 F_1=1.0

T_2=10305 K   q=0.885 F_2=1.0

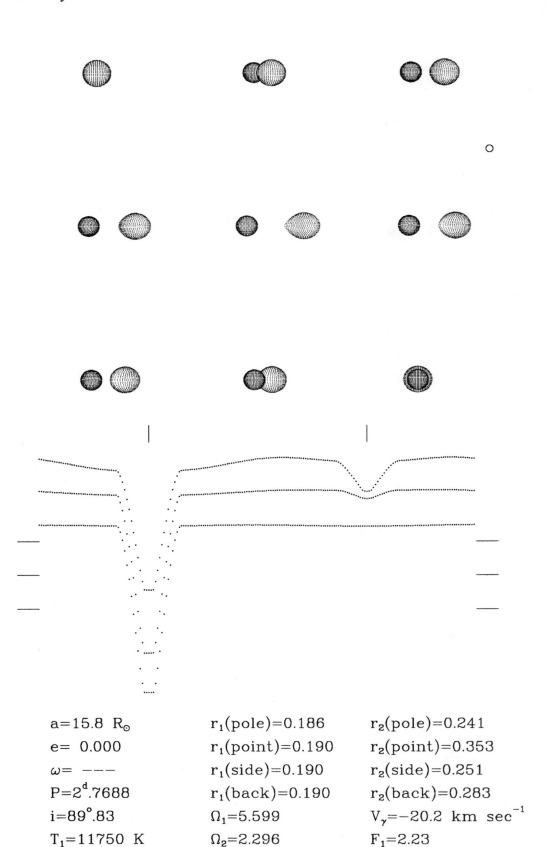

$$a=15.8 \ R_\odot \qquad r_1(\text{pole})=0.186 \qquad r_2(\text{pole})=0.241$$

$$e= \ 0.000 \qquad r_1(\text{point})=0.190 \qquad r_2(\text{point})=0.353$$

$$\omega= \ \text{---} \qquad r_1(\text{side})=0.190 \qquad r_2(\text{side})=0.251$$

$$P=2^d.7688 \qquad r_1(\text{back})=0.190 \qquad r_2(\text{back})=0.283$$

$$i=89^\circ.83 \qquad \Omega_1=5.599 \qquad V_\gamma=-20.2 \ \text{km sec}^{-1}$$

$$T_1=11750 \ K \qquad \Omega_2=2.296 \qquad F_1=2.23$$

$$T_2=4271 \ K \qquad q=0.2258 \qquad F_2=1.00$$

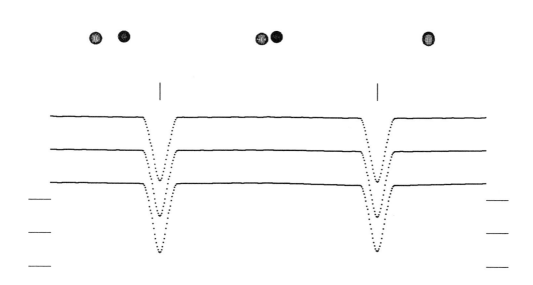

a= 13.35 $R_\odot$	r_1(pole)=0.129	r_2(pole)=0.123
e= 0.000	r_1(point)=0.129	r_2(point)=0.123
ω= –––	r_1(side)=0.129	r_2(side)=0.123
P=3^d.435	r_1(back)=0.129	r_2(back)=0.123
i=87°.28	Ω_1=8.685	V_γ=−29.5 km sec^{-1}
T_1=6580 K	Ω_2=8.583	F_1=1.00
T_2=6557 K	q=0.925	F_2=1.00

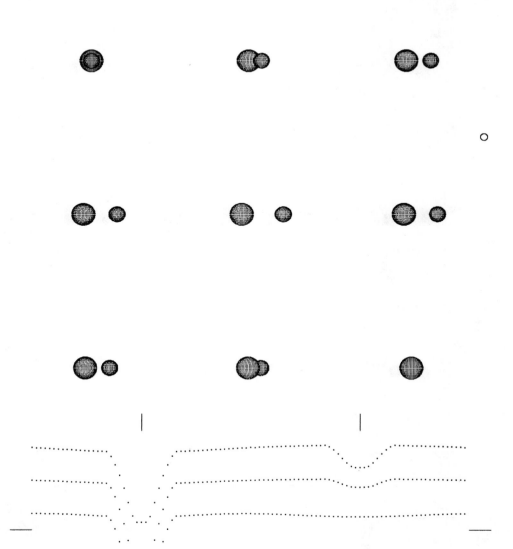

$a=11.51\ R_\odot$ $r_1(\text{pole})=0.281$ $r_2(\text{pole})=0.194$

$e=\ 0.000$ $r_1(\text{point})=0.289$ $r_2(\text{point})=0.207$

$\omega=\ ---$ $r_1(\text{side})=0.285$ $r_2(\text{side})=0.197$

$P=2^d.0563$ $r_1(\text{back})=0.288$ $r_2(\text{back})=0.204$

$i=90°.0$ $\Omega_1=3.850$ $V_\gamma=-2.3\ \text{km sec}^{-1}$

$T_1=12000\ K$ $\Omega_2=2.879$ $F_1=1.00$

$T_2=4970\ K$ $q=0.3$ $F_2=1.00$

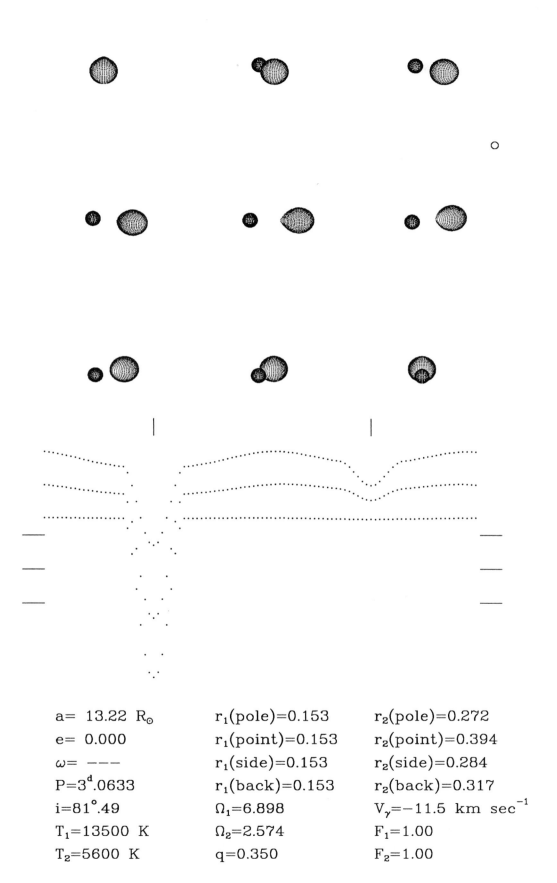

a= 13.22 $R_\odot$	r_1(pole)=0.153	r_2(pole)=0.272
e= 0.000	r_1(point)=0.153	r_2(point)=0.394
ω= ---	r_1(side)=0.153	r_2(side)=0.284
P=3^d.0633	r_1(back)=0.153	r_2(back)=0.317
i=81°.49	Ω_1=6.898	V_γ=−11.5 km sec^{-1}
T_1=13500 K	Ω_2=2.574	F_1=1.00
T_2=5600 K	q=0.350	F_2=1.00

a=11.77 $R_\odot$ r_1(pole)=0.101 r_2(pole)=0.106

e= 0.000 r_1(point)=0.101 r_2(point)=0.106

ω= --- r_1(side)=0.101 r_2(side)=0.106

P=4^d.6694 r_1(back)=0.101 r_2(back)=0.106

i=89°.66 Ω_1=10.929 V_γ=−29.8 km sec^{-1}

T_1=6400 K Ω_2=10.494 F_1=1.00

T_2=6350 K   q=1.004 F_2=1.00

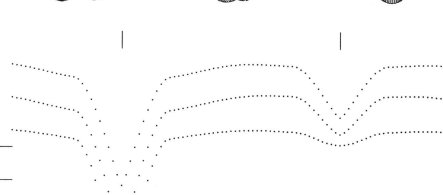

a= 15.02 $R_\odot$	r_1(pole)=0.303	r_2(pole)=0.288
e= 0.000	r_1(point)=0.317	r_2(point)=0.414
ω= ---	r_1(side)=0.309	r_2(side)=0.300
P=2^d.4549	r_1(back)=0.314	r_2(back)=0.333
i=88°.9	Ω_1=3.715	V_γ=−21.8 km sec^{-1}
T_1=19840 K	Ω_2=2.739	F_1=1.00
T_2=9410 K	q=0.43	F_2=1.00

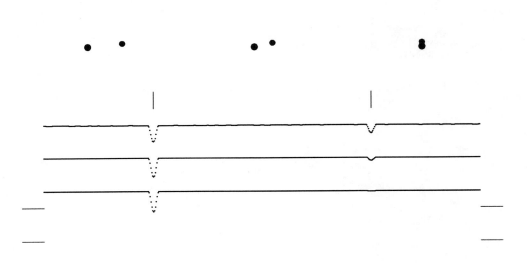

a= 16.31 $R_\odot$	r_1(pole)=0.055	r_2(pole)=0.048
e= 0.000	r_1(point)=0.055	r_2(point)=0.048
ω= ---	r_1(side)=0.055	r_2(side)=0.048
P=5^d.6092	r_1(back)=0.055	r_2(back)=0.048
i=85°.71	Ω_1=18.74	V_γ= ---
T_1=5470 K	Ω_2=16.12	F_1=1.00
T_2=3977 K	q=0.715	F_2=1.00

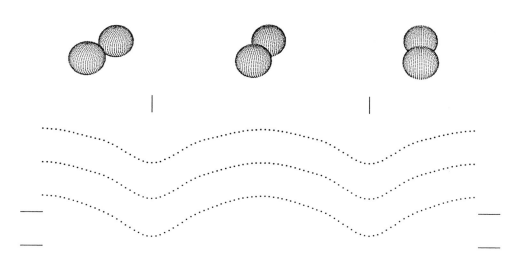

a=13.6 $R_\odot$	r_1(pole)=0.352	r_2(pole)=0.336
e= 0.000	r_1(point)=0.455	r_2(point)=0.400
ω= ———	r_1(side)=0.369	r_2(side)=0.350
P=1^d.2686	r_1(back)=0.398	r_2(back)=0.372
i=63°.24	Ω_1=3.785	V_γ= ———
T_1=24000 K	Ω_2=3.926	F_1=1.00
T_2=23451 K	q=1.0	F_2=1.00

Group VII

Systems with
$17.08\,R_\odot < a \leq 25.63\,R_\odot$

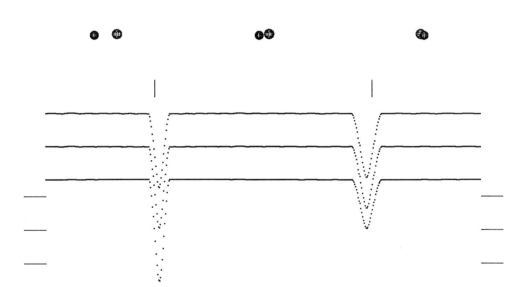

a= 21.24 R$_\odot$ r$_1$(pole)=0.085   r$_2$(pole)=0.098

e= 0.182   r$_1$(point)=0.085   r$_2$(point)=0.098

ω= 100°.81   r$_1$(side)=0.085   r$_2$(side)=0.098

P=6^d.7197   r$_1$(back)=0.085   r$_2$(back)=0.098

i=88°.69 Ω_1=12.906   V$_\gamma$=+9.9 km sec^{-1}

T$_1$=6500 K Ω_2=10.755   F$_1$=1.47

T$_2$=6095 K   q=0.93   F$_2$=1.47

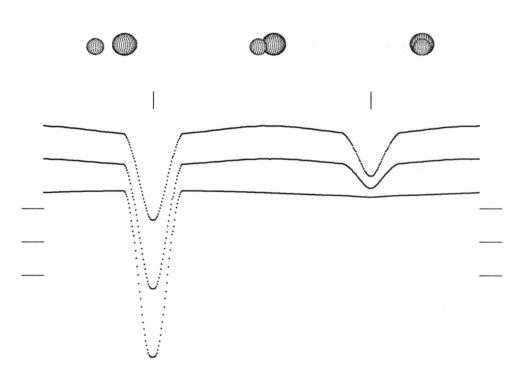

a= 20.71 R$_\odot$ r$_1$(pole)=0.175   r$_2$(pole)=0.236

e= 0.000   r$_1$(point)=0.176   r$_2$(point)=0.269

ω= ---   r$_1$(side)=0.175   r$_2$(side)=0.242

P=7^d.2296   r$_1$(back)=0.176   r$_2$(back)=0.258

i=86°.6 Ω_1=6.046   V$_\gamma$= +34 km sec^{-1}

T$_1$=7200 K Ω_2=2.709   F$_1$=1.00

T$_2$=4710 K   q=0.33   F$_2$=1.00

o

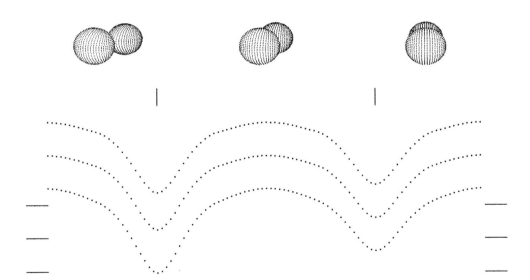

a= 21.5 R$_\odot$ r$_1$(pole)=0.374   r$_2$(pole)=0.314

e= 0.000   r$_1$(point)=0.441   r$_2$(point)=0.448

ω= ———   r$_1$(side)=0.392   r$_2$(side)=0.328

P=2^d.7339   r$_1$(back)=0.412   r$_2$(back)=0.361

i=80°.0 Ω_1=3.232   V$_\gamma$= ———

T$_1$=25000 K Ω_2=3.063   F$_1$=1.00

T$_2$=21800 K   q=0.6   F$_2$=1.00

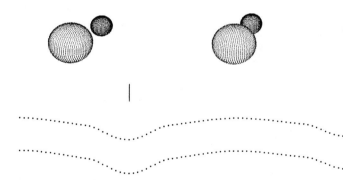

a= 21.05 R$_\odot$	r$_1$(pole)=0.413	r$_2$(pole)=0.209
e= 0.000	r$_1$(point)=0.478	r$_2$(point)=0.226
ω= ---	r$_1$(side)=0.436	r$_2$(side)=0.213
P=1^d.6219	r$_1$(back)=0.453	r$_2$(back)=0.222
i=63°.84	Ω_1=2.734	V$_\gamma$=+6.75 km sec^{-1}
T$_1$=36250 K	Ω_2=2.937	F$_1$=1.00
T$_2$=27500 K	q=0.34	F$_2$=1.00

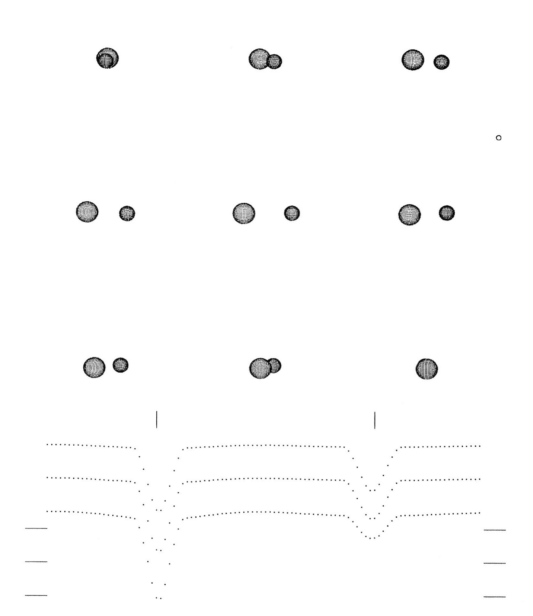

a= 20.47 $R_\odot$ r_1(pole)=0.220 r_2(pole)=0.157

e= 0.056 r_1(point)=0.228 r_2(point)=0.160

ω= 108°.5 r_1(side)=0.223 r_2(side)=0.158

P=3^d.1691 r_1(back)=0.226 r_2(back)=0.159

i=86°.4 Ω_1=5.422 V_γ=−4.3 km sec^{-1}

T_1=18740 K Ω_2=6.532 F_1=1.12

T_2=17790 K   q=0.85 F_2=1.12

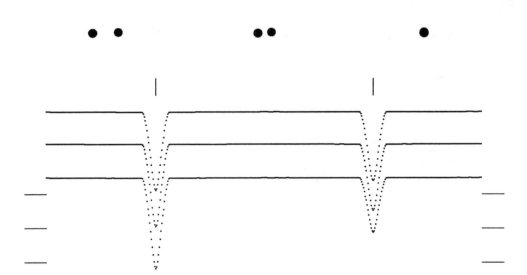

$$a=18.23 \; R_\odot \qquad r_1(pole)=0.100 \qquad r_2(pole)=0.096$$

$a=18.23 \; R_\odot$	$r_1(pole)=0.100$	$r_2(pole)=0.096$
$e= 0.000$	$r_1(point)=0.100$	$r_2(point)=0.096$
$\omega= \; ---$	$r_1(side)=0.100$	$r_2(side)=0.096$
$P=4^d.1346$	$r_1(back)=0.100$	$r_2(back)=0.096$
$i=88^\circ.6$	$\Omega_1=10.904$	$V_\gamma=+25.1 \; km \; sec^{-1}$
$T_1=11480 \; K$	$\Omega_2=10.426$	$F_1=1.00$
$T_2=10620 \; K$	$q=0.9$	$F_2=1.00$

a=23.34 $R_\odot$ r_1(pole)=0.044 r_2(pole)=0.038

e=0.003 r_1(point)=0.044 r_2(point)=0.038

ω=0°.0 r_1(side)=0.044 r_2(side)=0.038

P=9^d.8154 r_1(back)=0.044 r_2(back)=0.038

i=89°.79 Ω_1=23.708 V_γ=+27.1 km sec^{-1}

T_1=5346 K Ω_2=26.714 F_1=1.01

T_2=5200 K   q=0.977 F_2=1.01

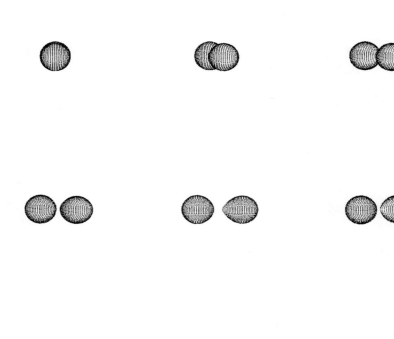

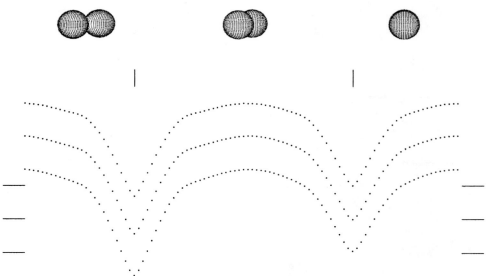

a=18.47 R$_\odot$ r$_1$(pole)=0.333   r$_2$(pole)=0.324

e= 0.000   r$_1$(point)=0.370   r$_2$(point)=0.460

ω= ---   r$_1$(side)=0.344   r$_2$(side)=0.339

P=1^د.8115   r$_1$(back)=0.358   r$_2$(back)=0.371

i=88°.2 Ω_1=3.645   V$_\gamma$=+13.0 km sec^{-1}

T$_1$=32000 K Ω_2=3.208   F$_1$=1.00

T$_2$=28560 K   q=0.68   F$_2$=1.00

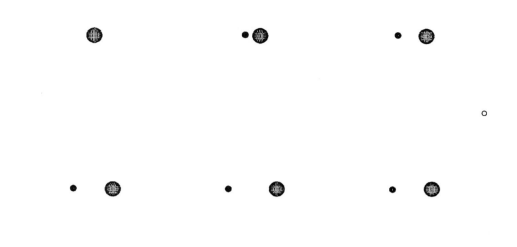

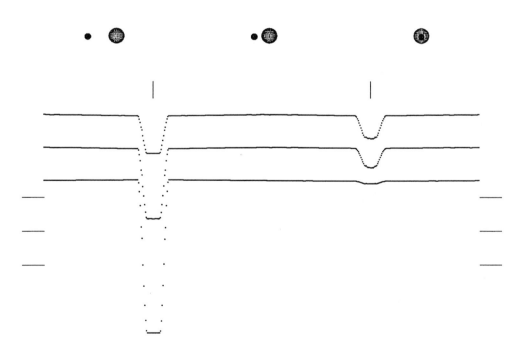

a= 20.28 $R_{\odot}$	r_1(pole)=0.064	r_2(pole)=0.159
e= 0.000	r_1(point)=0.064	r_2(point)=0.161
ω= ———	r_1(side)=0.064	r_2(side)=0.160
P=7^d.6061	r_1(back)=0.064	r_2(back)=0.161
i=$88°$.80	Ω_1=16.510	V_{γ}= −49 km sec^{-1}
T_1=5900 K	Ω_2=6.857	F_1=1.00
T_2=4400 K	q=1.004	F_2=1.00

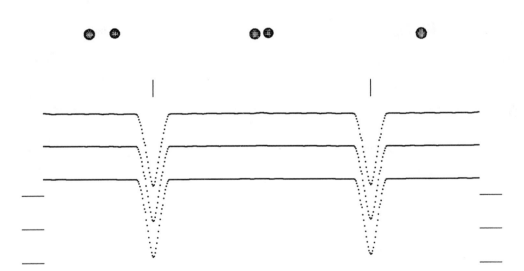

a= 18.07 $R_\odot$ r_1(pole)=0.125 r_2(pole)=0.115

e= 0.000 r_1(point)=0.125 r_2(point)=0.115

ω= ——— r_1(side)=0.125 r_2(side)=0.115

P=4^d.9917 r_1(back)=0.125 r_2(back)=0.115

i=88°.6 Ω_1=8.978 V_γ=−28.2 km sec^{-1}

T_1=6900 K Ω_2=9.445 F_1=1.00

T_2=6910 K   q=0.97 F_2=1.00

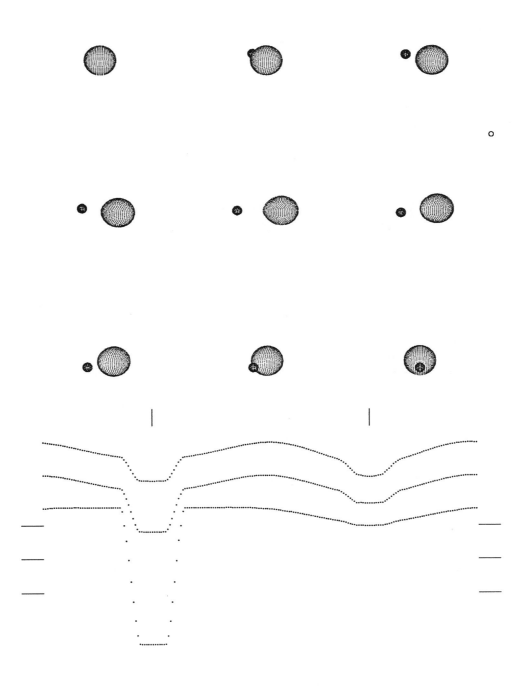

a=18.51 R$_\odot$	r$_1$(pole)=0.108	r$_2$(pole)=0.345
e= 0.000	r$_1$(point)=0.108	r$_2$(point)=0.419
ω= – – –	r$_1$(side)=0.108	r$_2$(side)=0.360
P=4^d.8242	r$_1$(back)=0.108	r$_2$(back)=0.384
i=80°.6	Ω_1=10.313	V$_\gamma$=−19.7 km sec^{-1}
T$_1$=6570 K	Ω_2=3.961	F$_1$=1.00
T$_2$=4770 K	q=1.048	F$_2$=1.00

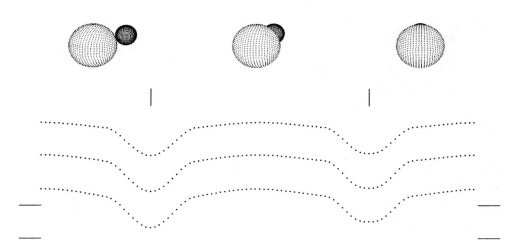

a=23.42 $R_\odot$ r_1(pole)=0.406 r_2(pole)=0.185

e= 0.000 r_1(point)=0.453 r_2(point)=0.197

ω= --- r_1(side)=0.426 r_2(side)=0.188

P=2^d.6985 r_1(back)=0.438 r_2(back)=0.195

i=$76°$.23 Ω_1=2.715 V_γ=−15.7 km sec^{-1}

T_1=30911 K Ω_2=2.824 F_1=1.00

T_2=28063 K   q=0.273 F_2=1.00

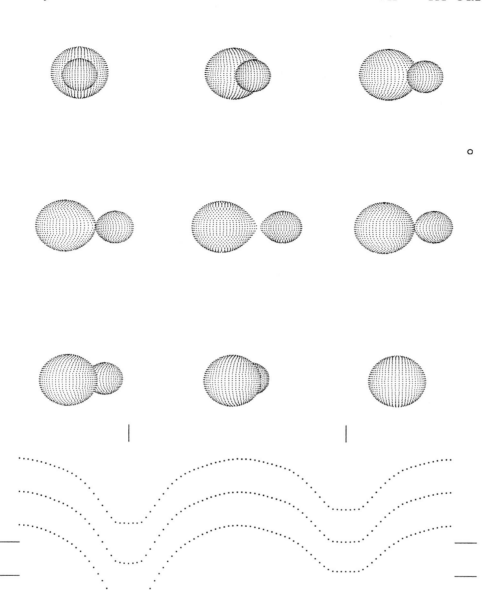

a=25.5 R$_\odot$	r$_1$(pole)=0.439	r$_2$(pole)=0.275
e= 0.000	r$_1$(point)=0.588	r$_2$(point)=0.368
ω= ---	r$_1$(side)=0.468	r$_2$(side)=0.286
P=2^d.9269	r$_1$(back)=0.495	r$_2$(back)=0.317
i=88°.0	Ω_1=2.620	V$_\gamma$= ---
T$_1$=33000 K	Ω_2=2.630	F$_1$=1.00
T$_2$=23720 K	q=0.371	F$_2$=1.00

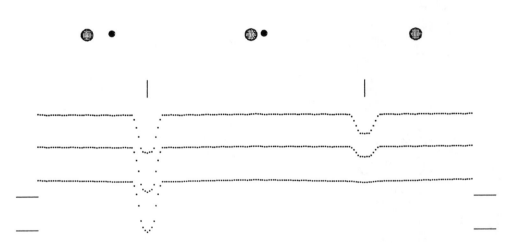

a= 17.57 $R_\odot$	$r_1(pole)=0.148$	$r_2(pole)=0.075$
e= 0.000	$r_1(point)=0.148$	$r_2(point)=0.075$
$\omega=$ ---	$r_1(side)=0.148$	$r_2(side)=0.075$
$P=4^d.4672$	$r_1(back)=0.148$	$r_2(back)=0.075$
$i=88^\circ.1$	$\Omega_1=7.351$	$V_\gamma=+8.1$ km sec^{-1}
$T_1=8800$ K	$\Omega_2=8.989$	$F_1=1.00$
$T_2=6000$ K	q=0.584	$F_2=1.00$

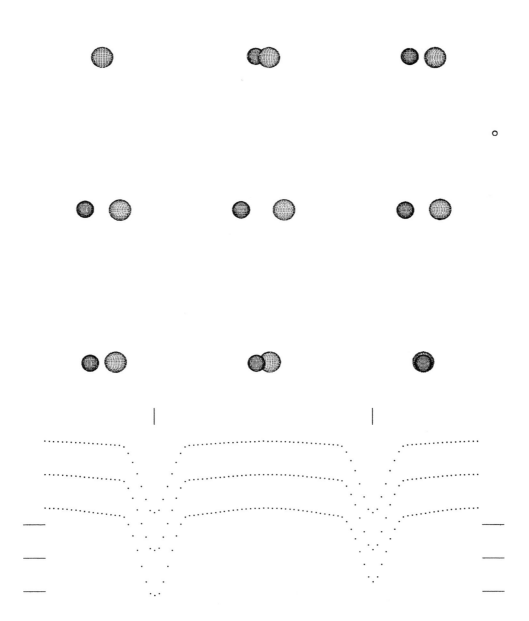

$$a=17.92 \ R_\odot \qquad r_1(\text{pole})=0.199 \qquad r_2(\text{pole})=0.249$$

$$e= \ 0.000 \qquad r_1(\text{point})=0.205 \qquad r_2(\text{point})=0.261$$

$$\omega= \ --- \qquad r_1(\text{side})=0.202 \qquad r_2(\text{side})=0.253$$

$$P=4^d.108 \qquad r_1(\text{back})=0.204 \qquad r_2(\text{back})=0.259$$

$$i=88°.24 \qquad \Omega_1=5.997 \qquad V_\gamma=-20.0 \ \text{km sec}^{-1}$$

$$T_1=8262 \ K \qquad \Omega_2=5.043 \qquad F_1=1.00$$

$$T_2=8000 \ K \qquad q=1.016 \qquad F_2=1.00$$

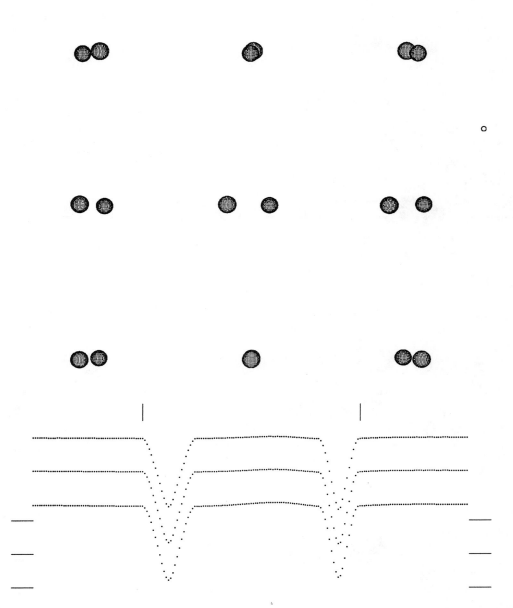

$a=22.0\ R_\odot$ $r_1(\text{pole})=0.169$ $r_2(\text{pole})=0.158$

$e=0.225$ $r_1(\text{point})=0.175$ $r_2(\text{point})=0.162$

$\omega=221°.0$ $r_1(\text{side})=0.171$ $r_2(\text{side})=0.159$

$P=4^d.1304$ $r_1(\text{back})=0.174$ $r_2(\text{back})=0.161$

$i=87°.5$ $\Omega_1=7.185$ $V_\gamma=\ ---$

$T_1=15000\ K$ $\Omega_2=7.620$ $F_1=1.62$

$T_2=14700\ K$ $q=1.0$ $F_2=1.62$

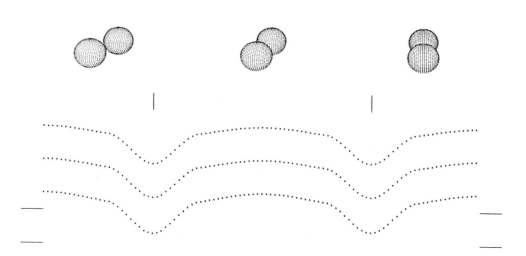

a= 19.85 $R_\odot$ r_1(pole)=0.322 r_2(pole)=0.299

e= 0.000 r_1(point)=0.362 r_2(point)=0.335

ω= - - - r_1(side)=0.333 r_2(side)=0.308

P=$1^d.7747$ r_1(back)=0.349 r_2(back)=0.324

i=$70^\circ.04$ Ω_1=3.942 V_γ=−21 km sec^{-1}

T_1=31500 K Ω_2=3.957 F_1=1.00

T_2=30377 K   q=0.88 F_2=1.00

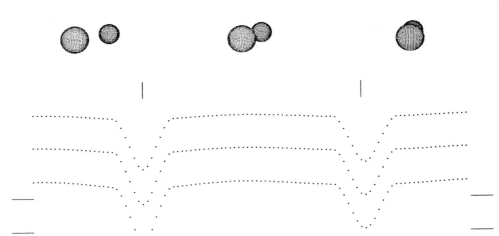

a= 23.07 $R_\odot$ r_1(pole)=0.246 r_2(pole)=0.181

e= 0.038 r_1(point)=0.259 r_2(point)=0.184

ω= 70°.20 r_1(side)=0.250 r_2(side)=0.182

P=2^d.7291 r_1(back)=0.256 r_2(back)=0.184

i=82°.85 Ω_1=5.010 V_γ= −13.9 km sec^{-1}

T_1=25412 K Ω_2=6.246 F_1=1.08

T_2=23539 K   q=0.939 F_2=1.08

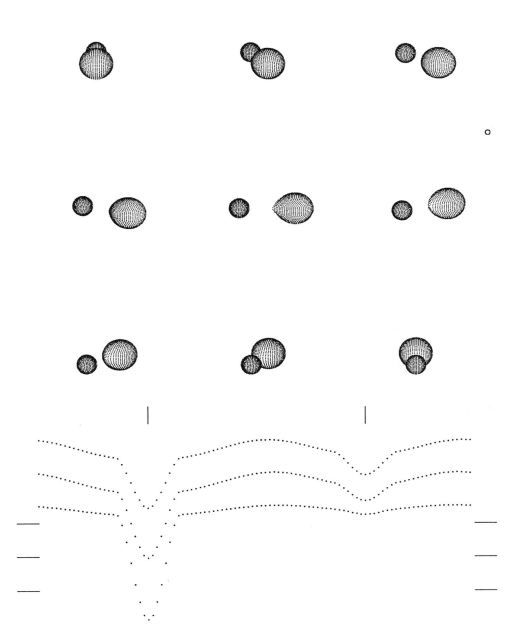

a= 23.27 R$_\odot$ r$_1$(pole)=0.180   r$_2$(pole)=0.288

e= 0.000   r$_1$(point)=0.181   r$_2$(point)=0.414

ω= −−−   r$_1$(side)=0.181   r$_2$(side)=0.300

P=4^d.9088   r$_1$(back)=0.181   r$_2$(back)=0.333

i=77°.3 Ω_1=5.983   V$_\gamma$= −4.9 km sec^{-1}

T$_1$=19000 K Ω_2=2.738   F$_1$=1.00

T$_2$=10145 K   q=0.43   F$_2$=1.00

$a=25.4\ R_{\odot}$ $r_1(\text{pole})=0.059$ $r_2(\text{pole})=0.050$

$e=0.025$ $r_1(\text{point})=0.059$ $r_2(\text{point})=0.050$

$\omega=114°.69$ $r_1(\text{side})=0.059$ $r_2(\text{side})=0.050$

$P=9^d.1033$ $r_1(\text{back})=0.059$ $r_2(\text{back})=0.050$

$i=89°.21$ $\Omega_1=18.121$ $V_\gamma=+21.5\ \text{km sec}^{-1}$

$T_1=7000\ K$ $\Omega_2=20.983$ $F_1=1.05$

$T_2=6593\ K$ $q=1.00$ $F_2=1.05$

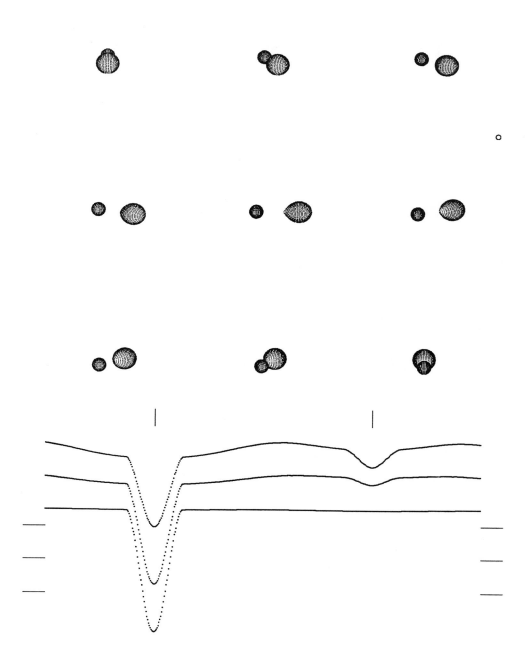

a= 17.85 $R_\odot$ r_1(pole)=0.161 r_2(pole)=0.279

e= 0.000 r_1(point)=0.162 r_2(point)=0.402

ω= — — — r_1(side)=0.162 r_2(side)=0.290

P=3^d.4522 r_1(back)=0.162 r_2(back)=0.229

i=$79°$.1 Ω_1=6.464 V_γ=−8.5 km sec^{-1}

T_1=15620 K Ω_2=2.422 F_1=1.00

T_2=5600 K   q=0.28 F_2=1.00

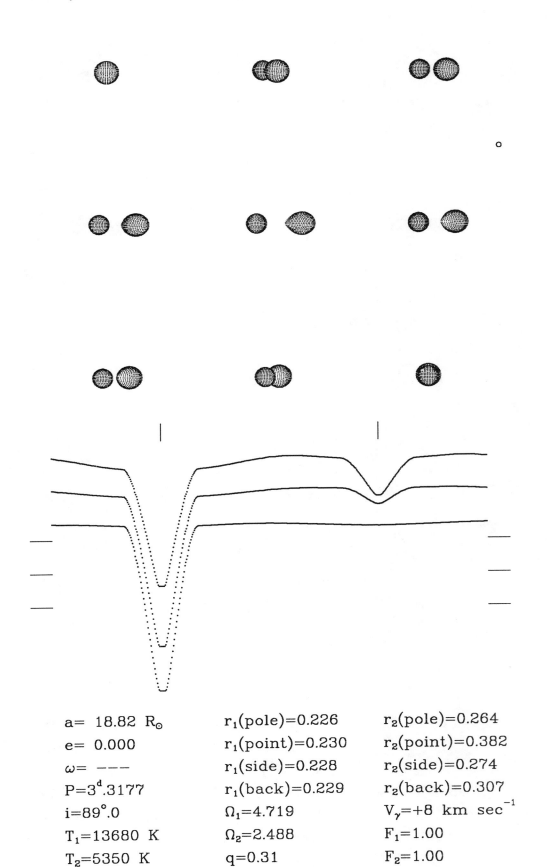

$a=$ 18.82 $R_\odot$ r_1(pole)=0.226 r_2(pole)=0.264

$e=$ 0.000 r_1(point)=0.230 r_2(point)=0.382

$\omega=$ — — — r_1(side)=0.228 r_2(side)=0.274

$P=3^d.3177$ r_1(back)=0.229 r_2(back)=0.307

$i=89°.0$ Ω_1=4.719 V_γ=+8 km sec^{-1}

T_1=13680 K Ω_2=2.488 F_1=1.00

T_2=5350 K q=0.31 F_2=1.00

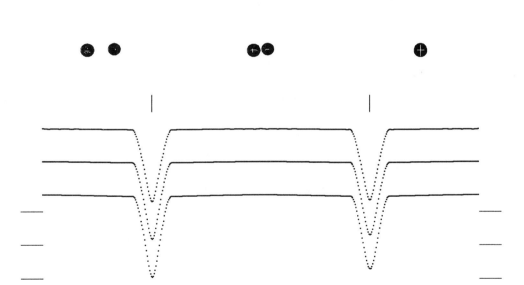

a= 18.97 R$_\odot$	r$_1$(pole)=0.144	r$_2$(pole)=0.133
e= 0.000	r$_1$(point)=0.145	r$_2$(point)=0.134
ω= ---	r$_1$(side)=0.144	r$_2$(side)=0.133
P=4^d.0052	r$_1$(back)=0.145	r$_2$(back)=0.134
i=88°.3	Ω_1=7.927	V$_\gamma$=−53.5 km sec^{-1}
T$_1$=8200 K	Ω_2=8.453	F$_1$=1.0
T$_2$=8140 K	q=0.99	F$_2$=1.0

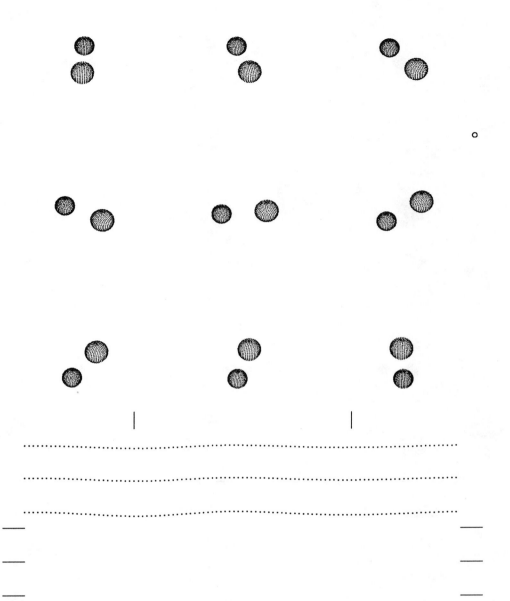

a= 19.34 $R_\odot$	r_1(pole)=0.209	r_2(pole)=0.243
e= 0.056	r_1(point)=0.215	r_2(point)=0.259
ω= 112°.7	r_1(side)=0.211	r_2(side)=0.248
P=1^d.8731	r_1(back)=0.214	r_2(back)=0.255
i=50°.00	Ω_1=5.695	V_γ= −8.6 km sec^{-1}
T_1=21100 K	Ω_2=4.707	F_1=1.12
T_2=19610 K	q=0.88	F_2=1.12

$a= 22.44 \ R_\odot$ $r_1(pole)=0.063$ $r_2(pole)=0.053$

$e= 0.5406$ $r_1(point)=0.064$ $r_2(point)=0.054$

$\omega= 48°.01$ $r_1(side)=0.064$ $r_2(side)=0.054$

$P=7^d.6408$ $r_1(back)=0.064$ $r_2(back)=0.054$

$i=87°.04$ $\Omega_1=17.879$ $V_\gamma=-16.10 \ km \ sec^{-1}$

$T_1=6550 \ K$ $\Omega_2=20.500$ $F_1=4.73$

$T_2=6496 \ K$ $q=0.979$ $F_2=4.73$

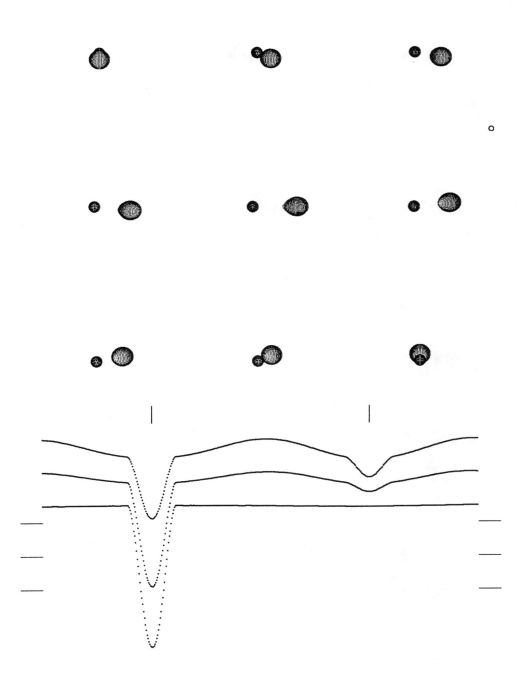

a= 18.36 R$_\odot$	r$_1$(pole)=0.129	r$_2$(pole)=0.226
e= 0.000	r$_1$(point)=0.129	r$_2$(point)=0.332
ω= ---	r$_1$(side)=0.129	r$_2$(side)=0.235
P=4^d.8060	r$_1$(back)=0.129	r$_2$(back)=0.267
i=84°.5	Ω_1=7.933	V$_\gamma$=+19 km sec^{-1}
T$_1$=9560 K	Ω_2=2.182	F$_1$=1.00
T$_2$=4910 K	q=0.18	F$_2$=1.00

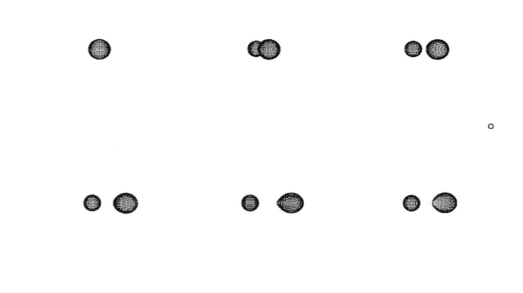

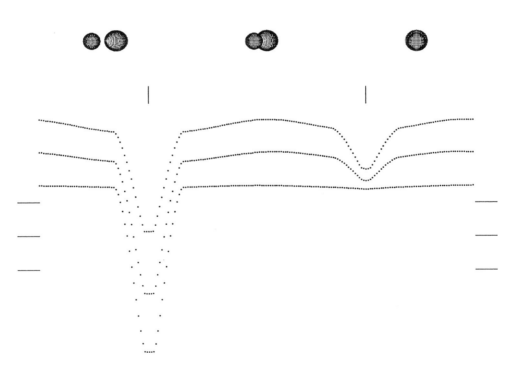

a=17.19 R$_\odot$	r$_1$(pole)=0.206	r$_2$(pole)=0.259
e= 0.000	r$_1$(point)=0.208	r$_2$(point)=0.376
ω= ---	r$_1$(side)=0.207	r$_2$(side)=0.269
P=2^d.8655	r$_1$(back)=0.208	r$_2$(back)=0.302
i=89°.2	Ω_1=5.142	V$_\gamma$=+0.3 km sec^{-1}
T$_1$=13010 K	Ω_2=2.444	F$_1$=1.00
T$_2$=6400 K	q=0.29	F$_2$=1.00

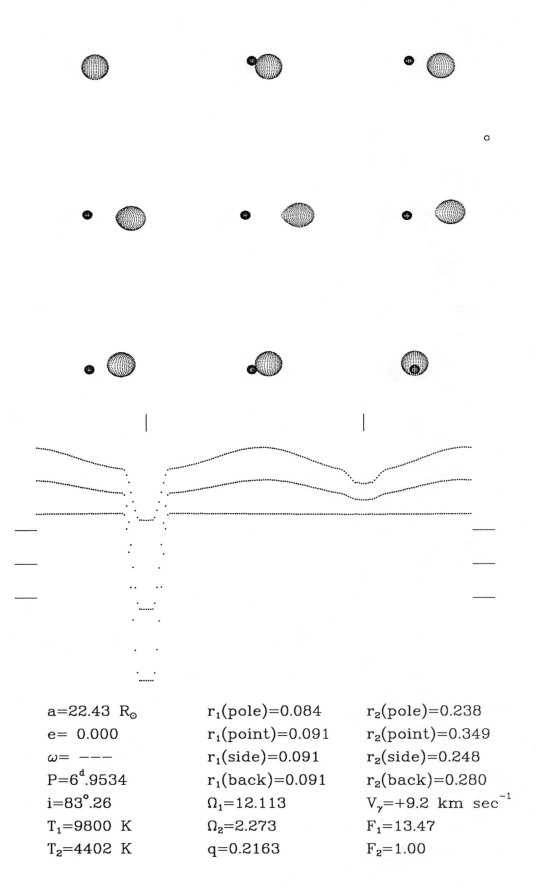

a=22.43 $R_\odot$ r_1(pole)=0.084 r_2(pole)=0.238

e= 0.000 r_1(point)=0.091 r_2(point)=0.349

ω= ——— r_1(side)=0.091 r_2(side)=0.248

P=6^d.9534 r_1(back)=0.091 r_2(back)=0.280

i=$83°$.26 Ω_1=12.113 V_γ=+9.2 km sec^{-1}

T_1=9800 K Ω_2=2.273 F_1=13.47

T_2=4402 K   q=0.2163 F_2=1.00

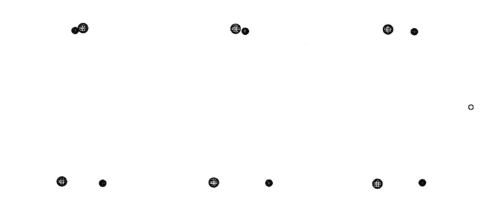

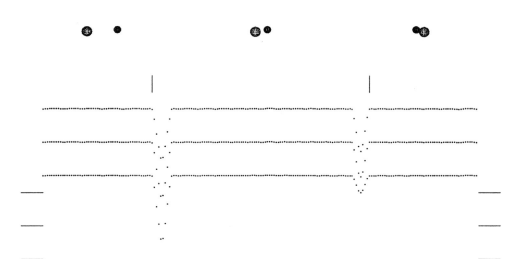

a=24.91 $R_\odot$ r_1(pole)=0.085 r_2(pole)=0.059

e=0.095 r_1(point)=0.085 r_2(point)=0.059

ω=225°.0 r_1(side)=0.085 r_2(side)=0.059

P=7^d.7505 r_1(back)=0.085 r_2(back)=0.059

i=87°.65 Ω_1=12.561 V_γ=−3.1 km sec^{-1}

T_1=8000 K Ω_2=13.906 F_1=1.21

T_2=6881 K   q=0.753 F_2=1.21

a=18.1 R$_\odot$ r$_1$(pole)=0.163 r$_2$(pole)=0.184

e= 0.000 r$_1$(point)=0.163 r$_2$(point)=0.274

ω= --- r$_1$(side)=0.163 r$_2$(side)=0.191

P=3^d.0199 r$_1$(back)=0.163 r$_2$(back)=0.223

i=86°.3 Ω_1=6.237 V$_\gamma$=+5.0 km sec^{-1}

T$_1$=9860 K Ω_2=1.928 F$_1$=1.00

T$_2$=5050 K q=0.09 F$_2$=1.00

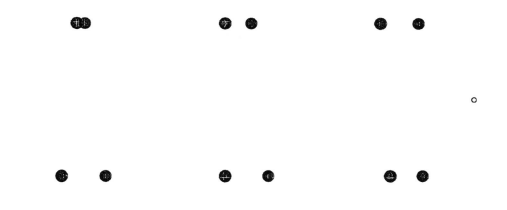

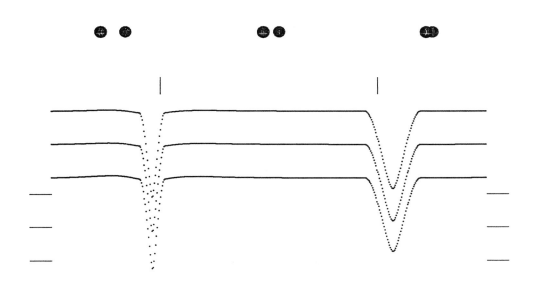

a=18.8 $R_\odot$ r_1(pole)=0.134 r_2(pole)=0.127

e=0.385 r_1(point)=0.139 r_2(point)=0.131

ω=78°.3 r_1(side)=0.136 r_2(side)=0.129

P=3^d.5847 r_1(back)=0.138 r_2(back)=0.131

i=89°.8 Ω_1=8.928 V_γ=+39.0 km sec^{-1}

T_1=12940 K Ω_2=8.895 F_1=2.44

T_2=12640 K   q=0.925 F_2=2.44

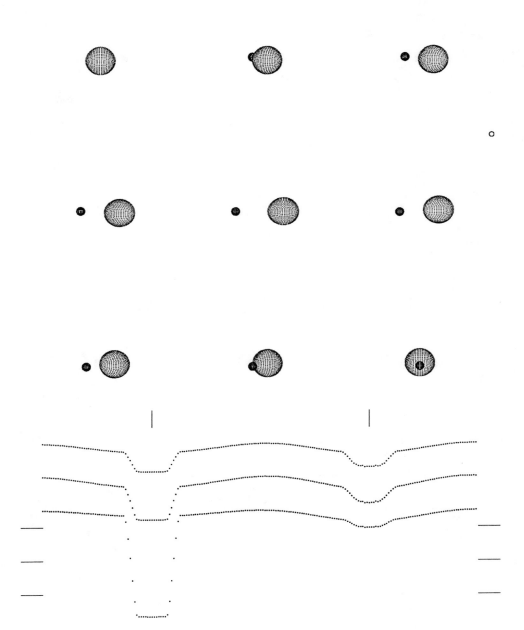

a=20.01 R$_\odot$ r$_1$(pole)=0.090   r$_2$(pole)=0.302

e= 0.000   r$_1$(point)=0.090   r$_2$(point)=0.332

ω= ---   r$_1$(side)=0.090   r$_2$(side)=0.310

P=6^d.0506   r$_1$(back)=0.090   r$_2$(back)=0.322

i=86°.0 Ω_1=12.175   V$_\gamma$=+19.5 km sec^{-1}

T$_1$=6400 K Ω_2=4.444   F$_1$=1.00

T$_2$=5000 K   q=1.06   F$_2$=1.00

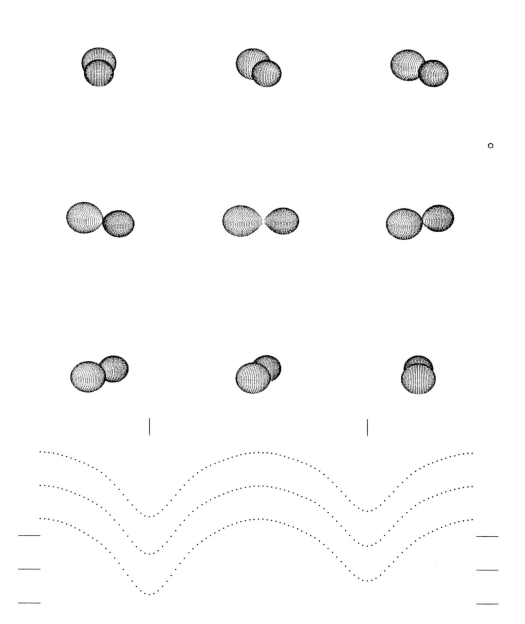

a=17.57 R$_\odot$ r$_1$(pole)=0.385   r$_2$(pole)=0.331

e= 0.000   r$_1$(point)=−1.000   r$_2$(point)=−1.000

ω= – – –   r$_1$(side)=0.407   r$_2$(side)=0.346

P=1^d.3873   r$_1$(back)=0.437   r$_2$(back)=0.379

i=75°.9 Ω_1=3.269   V$_\gamma$=+1.8 km sec^{-1}

T$_1$=33600 K Ω_2=3.269   F$_1$=1.00

T$_2$=29194 K   q=0.721   F$_2$=1.00

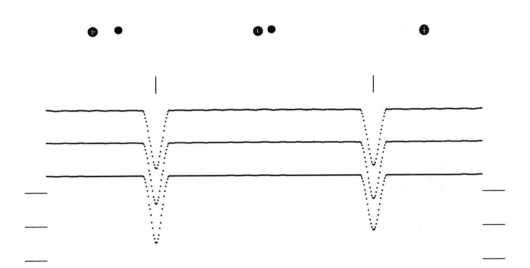

a= 18.24 R$_\odot$ r$_1$(pole)=0.108   r$_2$(pole)=0.086

e= 0.000   r$_1$(point)=0.108   r$_2$(point)=0.086

ω= –––   r$_1$(side)=0.108   r$_2$(side)=0.086

P=5^d.4899   r$_1$(back)=0.108   r$_2$(back)=0.086

i=87°.5 Ω_1=10.127   V$_\gamma$=−8.8 km sec^{-1}

T$_1$=6095 K Ω_2=11.370   F$_1$=1.0

T$_2$=6025 K   q=0.89   F$_2$=1.0

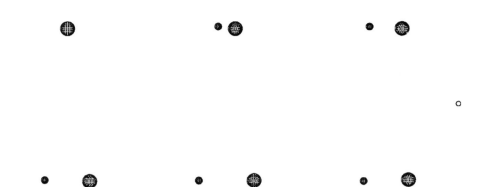

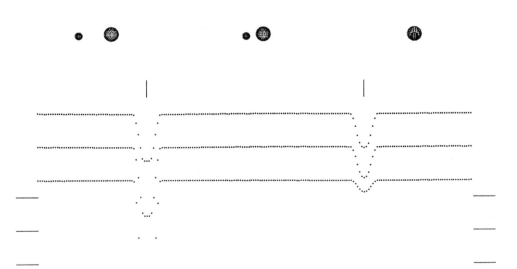

$a=23.10\ R_\odot$ $r_1(\text{pole})=0.067$ $r_2(\text{pole})=0.132$

$e=\ 0.000$ $r_1(\text{point})=0.067$ $r_2(\text{point})=0.133$

$\omega=\ ---$ $r_1(\text{side})=0.067$ $r_2(\text{side})=0.132$

$P=8^d.0382$ $r_1(\text{back})=0.067$ $r_2(\text{back})=0.132$

$i=87^\circ.58$ $\Omega_1=15.995$ $V_\gamma=+28.0\ \text{km sec}^{-1}$

$T_1=5800\ K$ $\Omega_2=9.059$ $F_1=1.00$

$T_2=4762\ K$ $q=1.067$ $F_2=1.00$

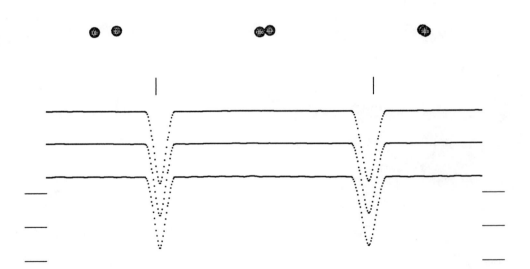

a=18.75 $R_\odot$ r_1(pole)=0.115 r_2(pole)=0.115

e=0.096 r_1(point)=0.116 r_2(point)=0.116

ω=108°.7 r_1(side)=0.116 r_2(side)=0.116

P=4^d.5962 r_1(back)=0.116 r_2(back)=0.116

i=88°.08 Ω_1=9.761 V_γ=+1.5 km sec^{-1}

T_1=9500 K Ω_2=9.760 F_1=1.22

T_2=9500 K   q=1.0 F_2=1.22

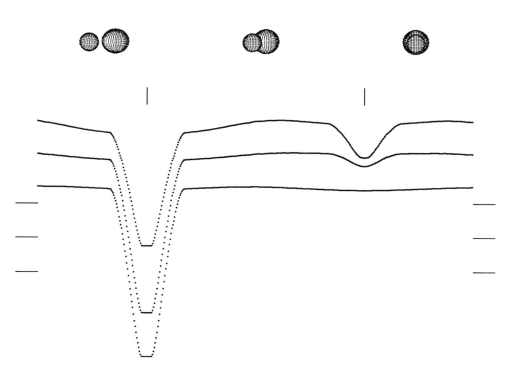

a= 18.49 $R_\odot$ r_1(pole)=0.213 r_2(pole)=0.283

e= 0.000 r_1(point)=0.218 r_2(point)=0.407

ω= ——— r_1(side)=0.216 r_2(side)=0.295

P=3^d.3806 r_1(back)=0.217 r_2(back)=0.328

i=$88°$.26 Ω_1=5.088 V_γ=−5.6 km sec^{-1}

T_1=12000 K Ω_2=2.681 F_1=1.44

T_2=4700 K   q=0.4013 F_2=1.00

o

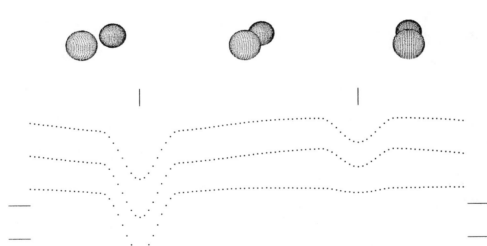

a=21.91 $R_\odot$ r_1(pole)=0.296 r_2(pole)=0.237

e= 0.000 r_1(point)=0.306 r_2(point)=0.286

ω= — — — r_1(side)=0.301 r_2(side)=0.245

P=3^d.9529 r_1(back)=0.304 r_2(back)=0.267

i=$76°$.2 Ω_1=3.638 V_γ=+13.8 km sec^{-1}

T_1=17960 K Ω_2=2.476 F_1=1.00

T_2=7710 K   q=0.27 F_2=1.00

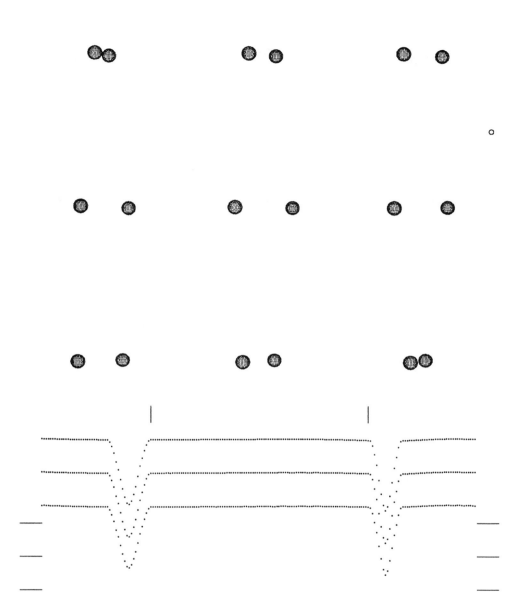

a=21.5 R$_\odot$ r$_1$(pole)=0.141   r$_2$(pole)=0.134

e=0.208   r$_1$(point)=0.142   r$_2$(point)=0.136

ω=313°.0   r$_1$(side)=0.141   r$_2$(side)=0.135

P=5^d.3297   r$_1$(back)=0.142   r$_2$(back)=0.136

i=86°.9 Ω_1=8.383   V$_\gamma$= ---

T$_1$=10500 K Ω_2=8.696   F$_1$=1.56

T$_2$=10500 K   q=1.0   F$_2$=1.56

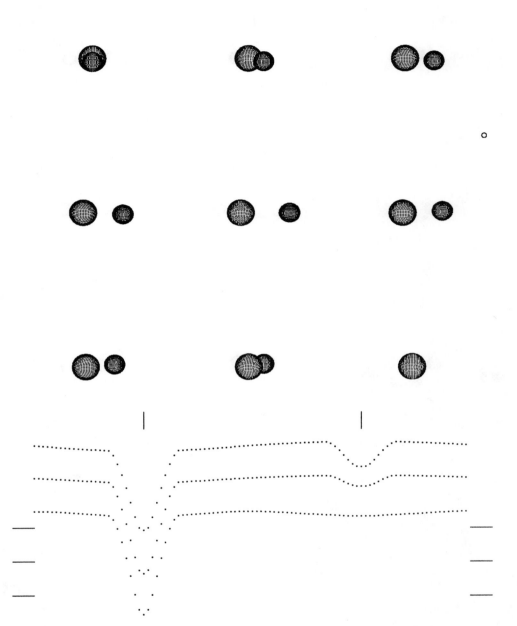

a= 20.43 R$_\odot$ r$_1$(pole)=0.275   r$_2$(pole)=0.203

e= 0.000   r$_1$(point)=0.283   r$_2$(point)=0.220

ω= — — —   r$_1$(side)=0.279   r$_2$(side)=0.207

P=4^d.4773   r$_1$(back)=0.282   r$_2$(back)=0.216

i=86°.3 Ω_1=3.933   V$_\gamma$=−22.2 km sec^{-1}

T$_1$=14950 K Ω_2=2.488   F$_1$=1.00

T$_2$=5490 K   q=0.31   F$_2$=1.00

Group VIII
Systems with
$25.63 \, R_\odot < a \leqslant 38.44 \, R_\odot$

a= 34.3 R$_\odot$ r$_1$(pole)=0.054   r$_2$(pole)=0.053

e= 0.37   r$_1$(point)=0.054   r$_2$(point)=0.053

ω= 130°.0   r$_1$(side)=0.054   r$_2$(side)=0.053

P=11^d.1209   r$_1$(back)=0.054   r$_2$(back)=0.053

i=88°.44 Ω_1=19.969   V$_\gamma$= ———

T$_1$=10200 K Ω_2=20.378   F$_1$=2.34

T$_2$=10500 K   q=1.0   F$_2$=2.34

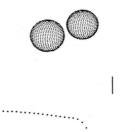

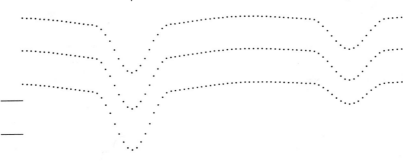

$a=37.35\ R_\odot$	$r_1(\text{pole})=0.278$	$r_2(\text{pole})=0.256$
$e=0.000$	$r_1(\text{point})=0.292$	$r_2(\text{point})=0.279$
$\omega=\ ---$	$r_1(\text{side})=0.283$	$r_2(\text{side})=0.262$
$P=4^d.0656$	$r_1(\text{back})=0.289$	$r_2(\text{back})=0.273$
$i=76°.6$	$\Omega_1=4.225$	$V_\gamma=-1.1\ \text{km sec}^{-1}$
$T_1=32000\ K$	$\Omega_2=3.680$	$F_1=1.00$
$T_2=20500\ K$	$q=0.65$	$F_2=1.00$

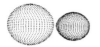 

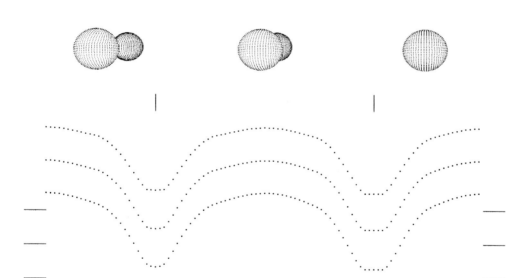

a=32.84 $R_\odot$	r_1(pole)=0.405	r_2(pole)=0.278
e= 0.000	r_1(point)=0.468	r_2(point)=0.399
ω= ---	r_1(side)=0.426	r_2(side)=0.289
P=4^d.0025	r_1(back)=0.443	r_2(back)=0.322
i=86°.9	Ω_1=2.817	V_γ=−2.5 km sec^{-1}
T_1=32020 K	Ω_2=2.627	F_1=1.00
T_2=32580 K	q=0.375	F_2=1.00

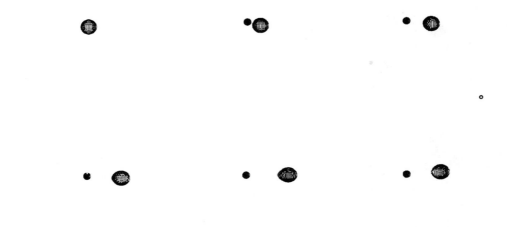

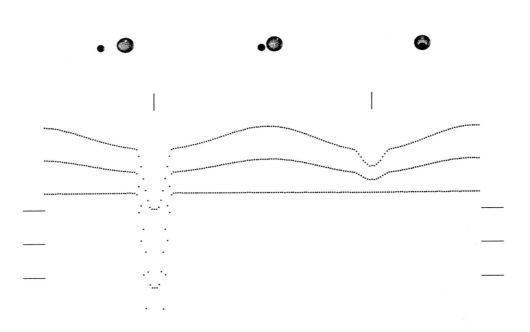

a=26.27 $R_\odot$ r_1(pole)=0.082 r_2(pole)=0.181

e= 0.000 r_1(point)=0.086 r_2(point)=0.270

ω= --- r_1(side)=0.086 r_2(side)=0.188

P=9^d.4845 r_1(back)=0.086 r_2(back)=0.218

i=84°.84 Ω_1=12.252 V_γ=+11.7 km sec^{-1}

T_1=10500 K Ω_2=1.912 F_1=10.6

T_2=4836 K   q=0.085 F_2=1.0

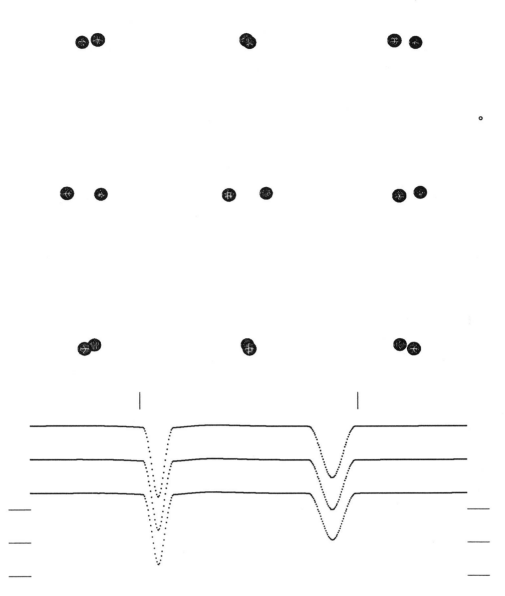

a=29.79 R$_\odot$ r$_1$(pole)=0.143   r$_2$(pole)=0.135

e=0.278   r$_1$(point)=0.146   r$_2$(point)=0.138

ω=123°.6   r$_1$(side)=0.144   r$_2$(side)=0.136

P=4^d.478   r$_1$(back)=0.145   r$_2$(back)=0.137

i=85°.7 Ω_1=8.255   V$_\gamma$=+17.0 km sec^{-1}

T$_1$=23800 K Ω_2=8.192   F$_1$=1.84

T$_2$=22600 K   q=0.915   F$_2$=1.84

 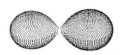 

a= 35.49 $R_\odot$	r_1(pole)=0.343	r_2(pole)=0.372
e= 0.000	r_1(point)=−1.000	r_2(point)=−1.000
ω= ---	r_1(side)=0.360	r_2(side)=0.392
P=3^d.5234	r_1(back)=0.392	r_2(back)=0.422
i=$51°$.2	Ω_1=4.034	V_γ=−31.1 km sec^{-1}
T_1=34700 K	Ω_2=4.034	F_1=1.00
T_2=35033 K	q=1.186	F_2=1.00

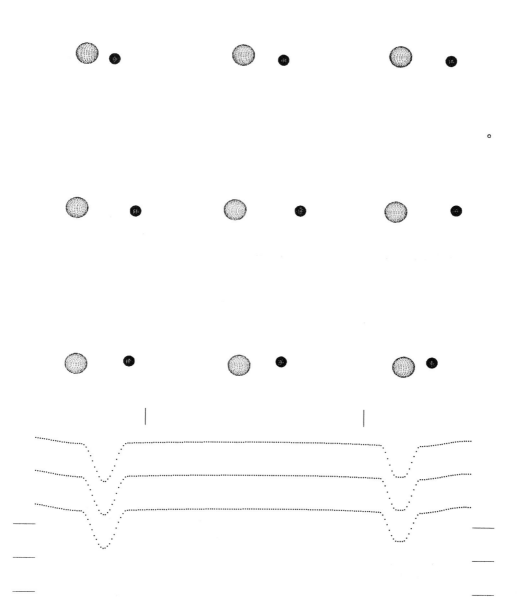

a=32.0 $R_\odot$ r_1(pole)=0.210 r_2(pole)=0.107

e=0.288 r_1(point)=0.226 r_2(point)=0.109

ω=345°.83 r_1(side)=0.216 r_2(side)=0.108

P=6^d.3219 r_1(back)=0.222 r_2(back)=0.109

i=83°.9 Ω_1=5.729 V_γ=−6.4 km sec^{-1}

T_1=26500 K Ω_2=8.160 F_1=1.89

T_2=24000 K   q=0.71 F_2=1.89

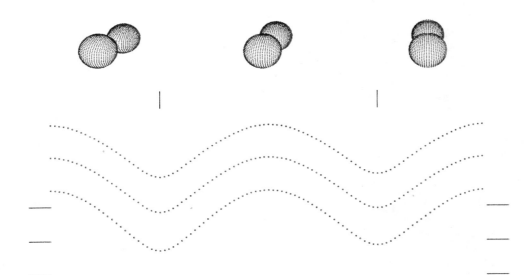

a= 26.3 $R_\odot$	r_1(pole)=0.418	r_2(pole)=0.360
e= 0.000	r_1(point)=−1.000	r_2(point)=−1.000
ω= ---	r_1(side)=0.448	r_2(side)=0.382
P=1^d.6412	r_1(back)=0.497	r_2(back)=0.441
i=68°.4	Ω_1=3.037	V_γ= +77 km sec^{-1}
T_1=45000 K	Ω_2=3.037	F_1=1.00
T_2=41750 K	q=0.7	F_2=1.00

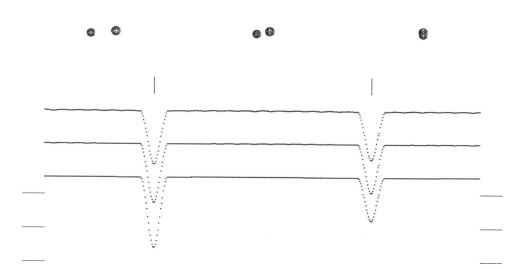

a= 26.40 $R_\odot$ r_1(pole)=0.097 r_2(pole)=0.108

e= 0.000 r_1(point)=0.097 r_2(point)=0.108

ω= --- r_1(side)=0.097 r_2(side)=0.108

P=8^d.4393 r_1(back)=0.097 r_2(back)=0.108

i=$86°.4$ Ω_1=11.338 V_γ=−1.1 km sec^{-1}

T_1=6700 K Ω_2=10.791 F_1=1.00

T_2=6400 K   q=1.06 F_2=1.00

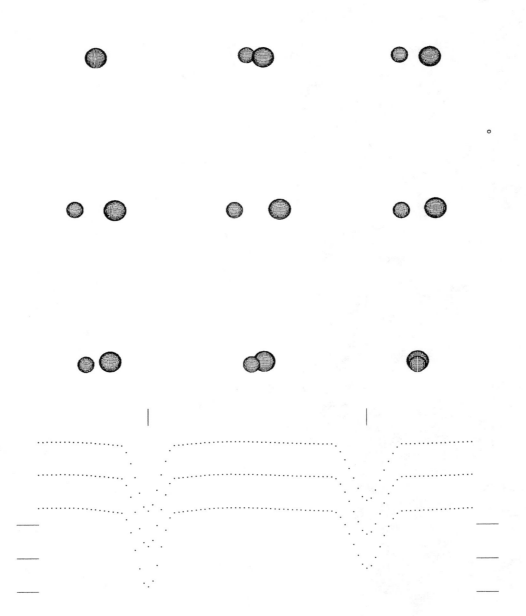

a= 28.67 R$_\odot$ r_1(pole)=0.179 r_2(pole)=0.230

e= 0.14 r_1(point)=0.185 r_2(point)=0.250

ω= 88°.7 r_1(side)=0.182 r_2(side)=0.238

P=2^d.9963 r_1(back)=0.184 r_2(back)=0.245

i=86°.4 Ω_1=6.721 V_γ=−57.5 km sec^{-1}

T$_1$=29600 K Ω_2=5.473 F_1=1.64

T$_2$=26950 K q=1.0 F_2=1.64

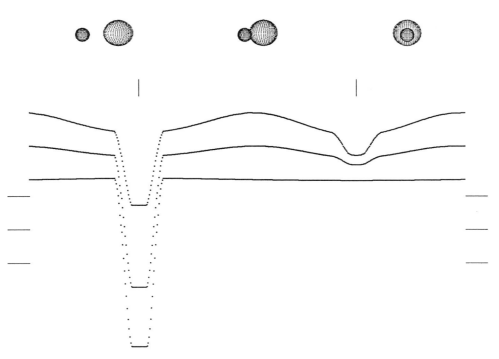

a= 37.82 R$_\odot$	r$_1$(pole)=0.117	r$_2$(pole)=0.223
e= 0.000	r$_1$(point)=0.117	r$_2$(point)=0.310
ω= −−−	r$_1$(side)=0.117	r$_2$(side)=0.232
P=8^d.4303	r$_1$(back)=0.117	r$_2$(back)=0.264
i=88°.02	Ω_1=8.725	V$_\gamma$= −30 km sec^{-1}
T$_1$=8840 K	Ω_2=2.164	F$_1$=1.00
T$_2$=4336 K	q=0.173	F$_2$=1.00

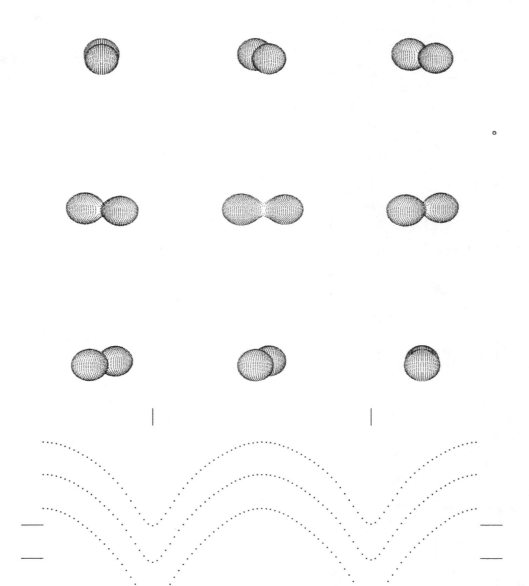

a= 26.58 $R_\odot$ r_1(pole)=0.393 r_2(pole)=0.371

e= 0.000 r_1(point)=−1.000 r_2(point)=−1.000

ω= − − − r_1(side)=0.419 r_2(side)=0.395

P=1^d.8855 r_1(back)=0.465 r_2(back)=0.444

i=83°.7 Ω_1=3.363 V_γ=+9.2 km sec^{-1}

T_1=36010 K Ω_2=3.363 F_1=1.0

T_2=34950 K   q=0.88 F_2=1.0

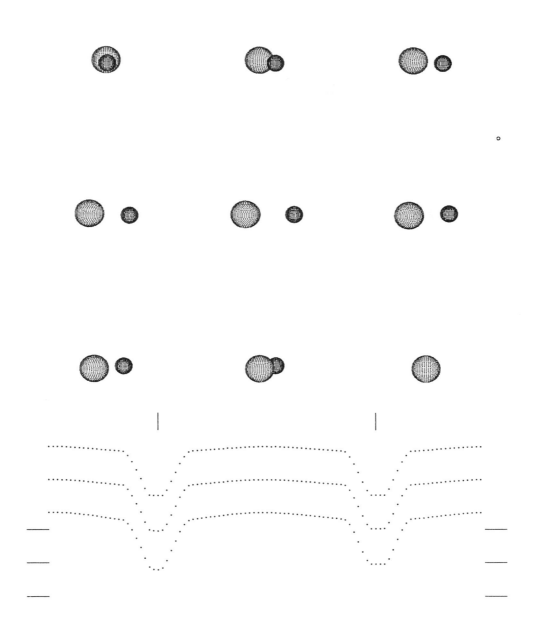

a= 30.27 $R_\odot$	r_1(pole)=0.289	r_2(pole)=0.176
e= 0.012	r_1(point)=0.310	r_2(point)=0.179
ω= 0°.0	r_1(side)=0.296	r_2(side)=0.177
P=3^d.8898	r_1(back)=0.304	r_2(back)=0.179
i=86°.1	Ω_1=4.206	V_γ= −14 km sec^{-1}
T_1=28200 K	Ω_2=5.490	F_1=1.02
T_2=27910 K	q=0.77	F_2=1.02

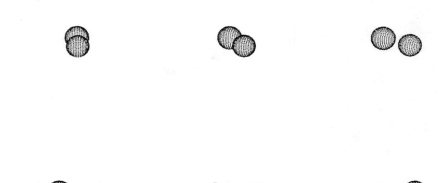

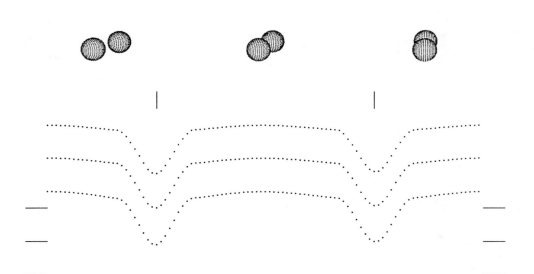

a= 27.28 $R_\odot$ r_1(pole)=0.271 r_2(pole)=0.259

e= 0.015 r_1(point)=0.290 r_2(point)=0.275

ω= 50°.8 r_1(side)=0.277 r_2(side)=0.264

P=2^d.8808 r_1(back)=0.285 r_2(back)=0.271

i=78°.07 Ω_1=4.648 V_γ= −15 km sec^{-1}

T_1=30900 K Ω_2=4.776 F_1=1.03

T_2=30308 K   q=0.98 F_2=1.03

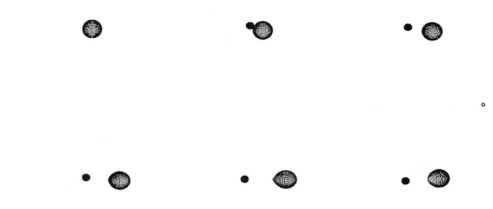

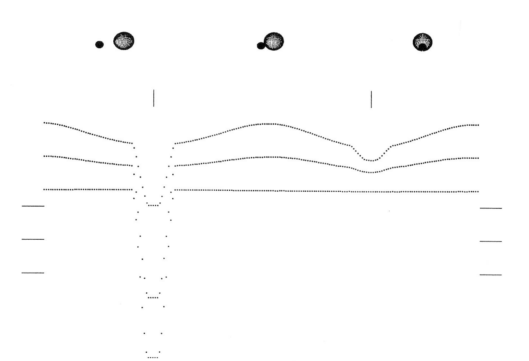

a= 25.79 R$_\odot$	r$_1$(pole)=0.086	r$_2$(pole)=0.227
e= 0.000	r$_1$(point)=0.096	r$_2$(point)=0.333
ω= ---	r$_1$(side)=0.096	r$_2$(side)=0.236
P=9^d.3005	r$_1$(back)=0.096	r$_2$(back)=0.268
i=83°.10	Ω_1=11.768	V$_\gamma$=+15.1 km sec^{-1}
T$_1$=9400 K	Ω_2=2.187	F$_1$=14.42
T$_2$=4043 K	q=0.1819	F$_2$=1.00

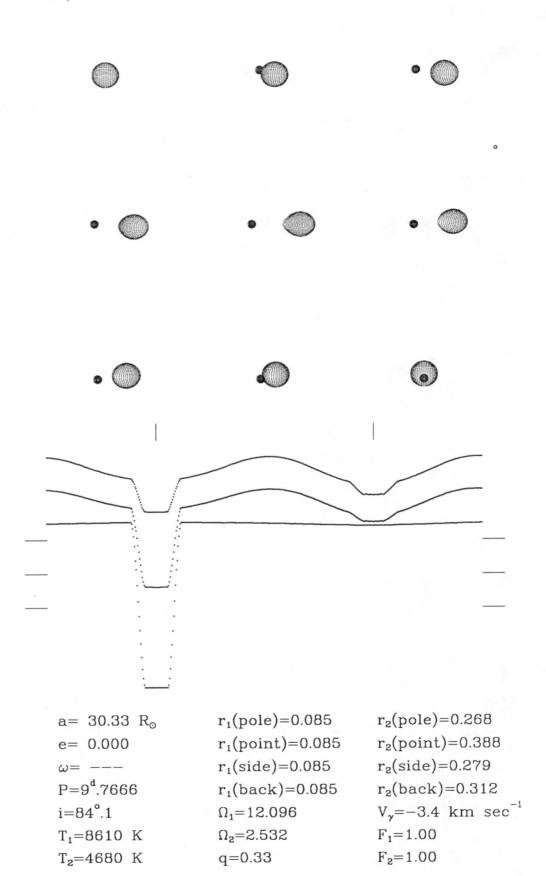

a= 30.33 $R_\odot$ r_1(pole)=0.085 r_2(pole)=0.268

e= 0.000 r_1(point)=0.085 r_2(point)=0.388

ω= ——— r_1(side)=0.085 r_2(side)=0.279

P=9^d.7666 r_1(back)=0.085 r_2(back)=0.312

i=$84°$.1 Ω_1=12.096 V_γ=−3.4 km sec^{-1}

T_1=8610 K Ω_2=2.532 F_1=1.00

T_2=4680 K   q=0.33 F_2=1.00

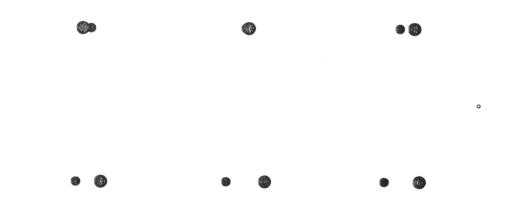

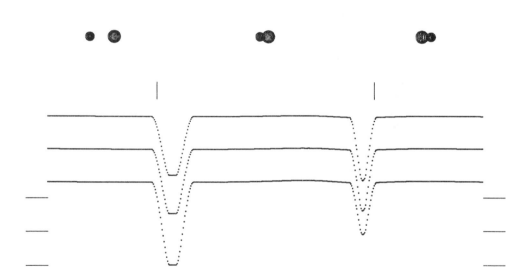

a=28.4 R$_\odot$	r$_1$(pole)=0.100	r$_2$(pole)=0.141
e=0.232	r$_1$(point)=0.101	r$_2$(point)=0.143
ω=245°.82	r$_1$(side)=0.100	r$_2$(side)=0.142
P=8^d.2897	r$_1$(back)=0.100	r$_2$(back)=0.143
i=89°.98	Ω_1=11.416	V$_\gamma$=+45.1 km sec^{-1}
T$_1$=8040 K	Ω_2=8.939	F$_1$=1.65
T$_2$=7500 K	q=1.084	F$_2$=1.65

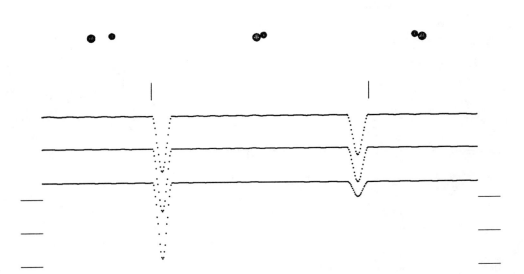

a=30.08 $R_\odot$ r_1(pole)=0.085 r_2(pole)=0.067

e=0.09 r_1(point)=0.085 r_2(point)=0.067

ω=150°.1 r_1(side)=0.085 r_2(side)=0.067

P=9^d.9451 r_1(back)=0.085 r_2(back)=0.067

i=87°.26 Ω_1=12.622 V_γ=−13.2 km sec^{-1}

T_1=8620 K Ω_2=12.370 F_1=1.20

T_2=7240 K   q=0.75 F_2=1.20

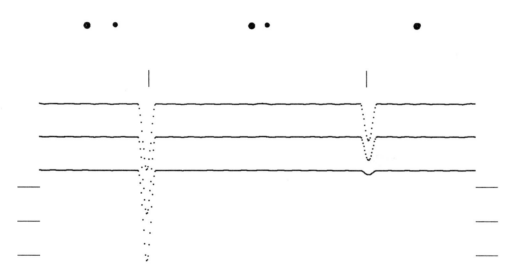

a=27.95 $R_\odot$ r_1(pole)=0.072 r_2(pole)=0.051

e=0.035 r_1(point)=0.072 r_2(point)=0.051

ω=295°.1 r_1(side)=0.072 r_2(side)=0.051

P=8^d.569 r_1(back)=0.072 r_2(back)=0.051

i=88°.73 Ω_1=14.481 V_γ=−0.3 km sec^{-1}

T_1=10400 K Ω_2=13.024 F_1=1.07

T_2=7200 K   q=0.605 F_2=1.07

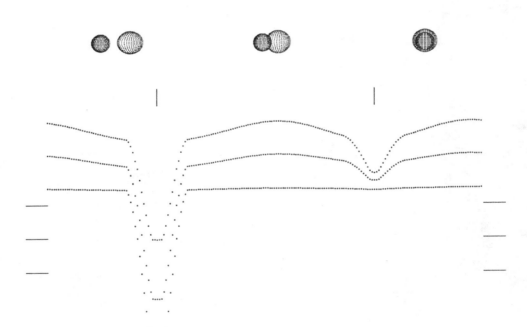

a= 31.43 R$_\odot$	r$_1$(pole)=0.183	r$_2$(pole)=0.236
e= 0.000	r$_1$(point)=0.184	r$_2$(point)=0.346
ω= ---	r$_1$(side)=0.184	r$_2$(side)=0.246
P=5^d.049	r$_1$(back)=0.184	r$_2$(back)=0.278
i=89°.1	Ω_1=5.662	V$_\gamma$=+20.0 km sec^{-1}
T$_1$=15520 K	Ω_2=2.257	F$_1$=1.00
T$_2$=7280 K	q=0.21	F$_2$=1.00

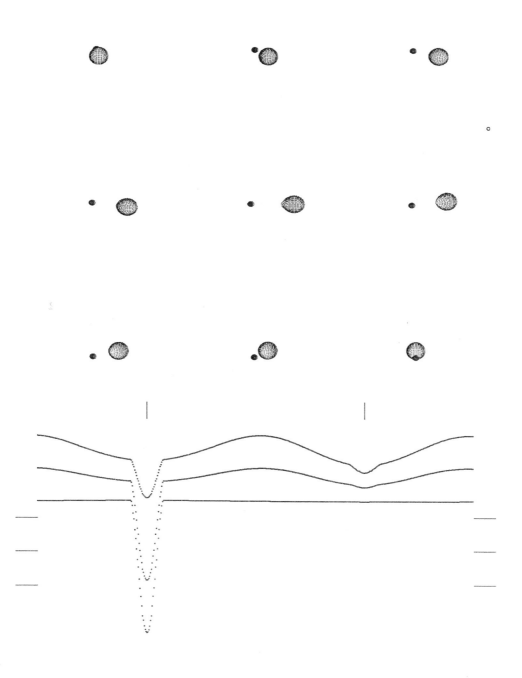

a= 26.96 $R_\odot$ r_1(pole)=0.062 r_2(pole)=0.209

e= 0.000 r_1(point)=0.076 r_2(point)=0.309

ω= − − − r_1(side)=0.076 r_2(side)=0.218

P=10^d.622 r_1(back)=0.076 r_2(back)=0.249

i=80°.30 Ω_1=16.160 V_γ=−12.4 km sec^{-1}

T_1=9000 K Ω_2=2.072 F_1=29.30

T_2=3928 K   q=0.1387 F_2=1.00

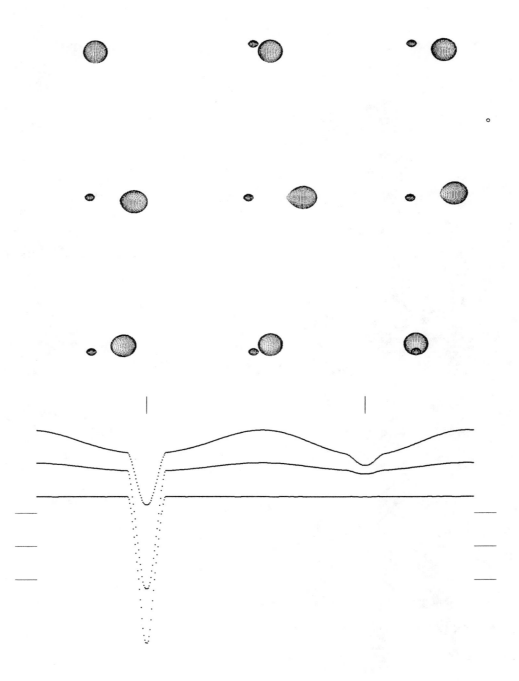

a=34.10 R$_\odot$ r$_1$(pole)=0.067   r$_2$(pole)=0.214

e= 0.000   r$_1$(point)=0.094   r$_2$(point)=0.315

ω= ———   r$_1$(side)=0.094   r$_2$(side)=0.223

P=13^d.1989   r$_1$(back)=0.094   r$_2$(back)=0.255

i=81°.56 Ω_1=14.999   V$_\gamma$=+6.5 km sec^{-1}

T$_1$=9700 K Ω_2=2.104   F$_1$=28.80

T$_2$=3853 K   q=0.15   F$_2$=1.00

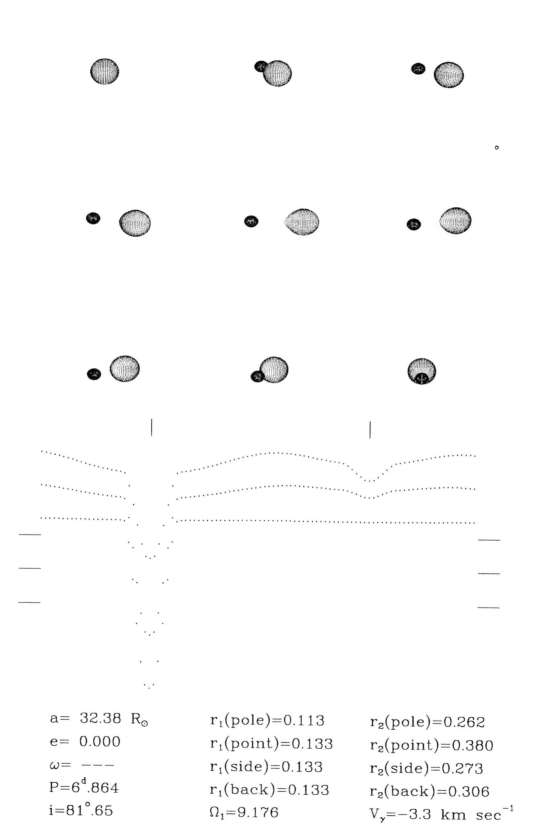

a= 32.38 R$_\odot$ r$_1$(pole)=0.113   r$_2$(pole)=0.262

e= 0.000   r$_1$(point)=0.133   r$_2$(point)=0.380

ω= - - -   r$_1$(side)=0.133   r$_2$(side)=0.273

P=6^{d}.864   r$_1$(back)=0.133   r$_2$(back)=0.306

i=81°.65 Ω_1=9.176   V$_\gamma$=−3.3 km sec^{-1}

T$_1$=20700 K Ω_2=2.474   F$_1$=10.82

T$_2$=6017 K   q=0.3037   F$_2$=1.00

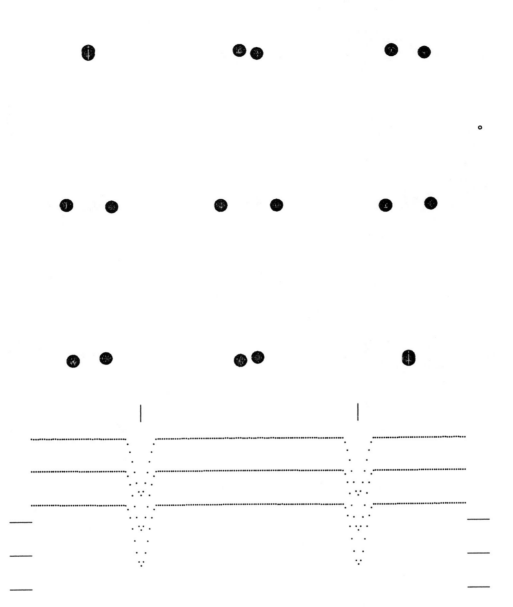

a=34.94 R$_\odot$ r$_1$(pole)=0.117   r$_2$(pole)=0.113

e= 0.000   r$_1$(point)=0.117   r$_2$(point)=0.113

ω= ———   r$_1$(side)=0.117   r$_2$(side)=0.113

P=6^d.8895   r$_1$(back)=0.117   r$_2$(back)=0.113

i=86°.59 Ω_1=9.537   V$_\gamma$=+24.3 km sec^{-1}

T$_1$=18200 K Ω_2=9.713   F$_1$=1.00

T$_2$=18060 K   q=0.983   F$_2$=1.00

a= 27.5 $R_\odot$	r_1(pole)=0.562	r_2(pole)= ———
e= 0.000	r_1(point)=0.753	r_2(point)= ———
ω= ———	r_1(side)=0.639	r_2(side)= ———
P=3^d.8923	r_1(back)=0.654	r_2(back)= ———
i=64°.0	Ω_1=1.835	V_γ= +180 km sec^{-1}
T_1=31500 K	Ω_2= ———	F_1=1.00
T_2= ———	q=0.0625	F_2=1.00

Group IX
Systems with
$38.44\ R_\odot < a \leqslant 57.37\ R_\odot$

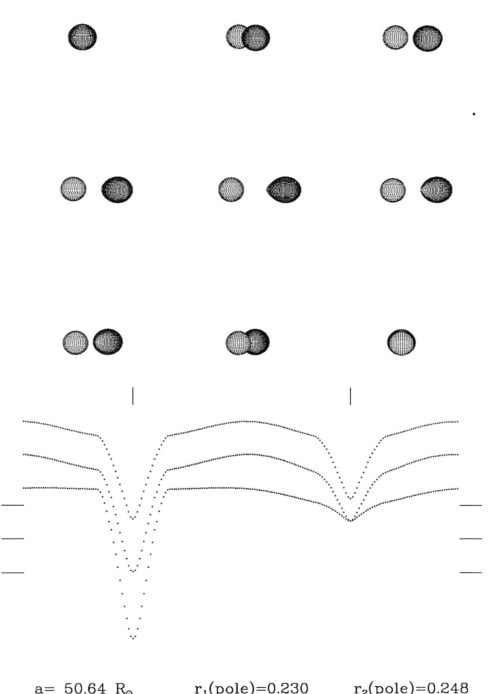

a= 50.64 R$_\odot$	r$_1$(pole)=0.230	r$_2$(pole)=0.248
e= 0.000	r$_1$(point)=0.233	r$_2$(point)=0.359
ω= ---	r$_1$(side)=0.231	r$_2$(side)=0.258
P=21^d.643	r$_1$(back)=0.232	r$_2$(back)=0.291
i=88°.7	Ω_1=4.60	V$_\gamma$=+14.2 km sec^{-1}
T$_1$=4780 K	Ω_2=2.35	F$_1$=1.00
T$_2$=4062 K	q=0.25	F$_2$=1.00

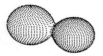 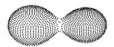 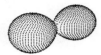

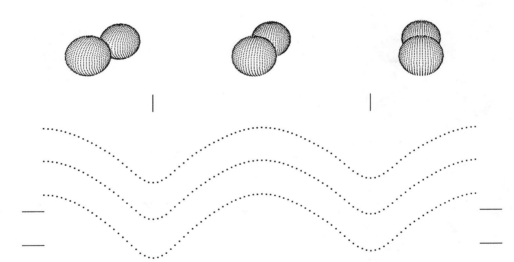

a=49.6 $R_\odot$ r_1(pole)=0.395 r_2(pole)=0.347

e= 0.000 r_1(point)=−1.000 r_2(point)=−1.000

ω= −−− r_1(side)=0.419 r_2(side)=0.366

P=4^d.3934 r_1(back)=0.457 r_2(back)=0.407

i=70°.21 Ω_1=3.231 V_γ=+20.0 km sec^{-1}

T_1=43000 K Ω_2=3.231 F_1=1.00

T_2=39194 K   q=0.751 F_2=1.00

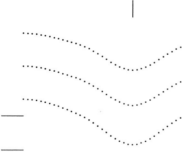 

$a=52.27\ R_\odot$	$r_1(\text{pole})=0.344$	$r_2(\text{pole})=0.360$
$e=0.000$	$r_1(\text{point})=0.445$	$r_2(\text{point})=0.465$
$\omega=\ ---$	$r_1(\text{side})=0.360$	$r_2(\text{side})=0.378$
$P=5^d.5621$	$r_1(\text{back})=0.389$	$r_2(\text{back})=0.407$
$i=63^\circ.0$	$\Omega_1=3.947$	$V_\gamma=+0.6\ \text{km sec}^{-1}$
$T_1=29700\ K$	$\Omega_2=3.947$	$F_1=1.00$
$T_2=26200\ K$	$q=1.1$	$F_2=1.00$

a= 42.92 R$_\odot$ r$_1$(pole)=0.077   r$_2$(pole)=0.022

e= 0.000   r$_1$(point)=0.077   r$_2$(point)=0.022

ω= ———   r$_1$(side)=0.077   r$_2$(side)=0.022

P=17^d.3599   r$_1$(back)=0.077   r$_2$(back)=0.022

i=87°.77 Ω_1=13.500   V$_\gamma$=+1.4 km sec^{-1}

T$_1$=9570 K Ω_2=18.096   F$_1$=1.00

T$_2$=5000 K   q=0.36   F$_2$=1.00

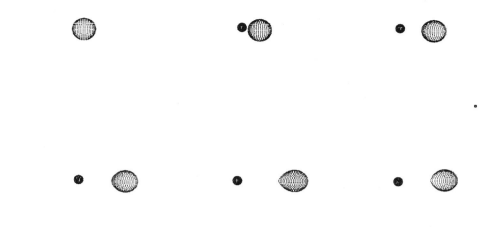

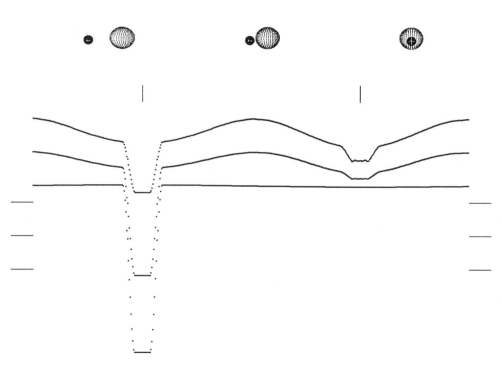

a= 53.4 R$_\odot$ r$_1$(pole)=0.079   r$_2$(pole)=0.173

e= 0.000   r$_1$(point)=0.079   r$_2$(point)=0.198

ω= ———   r$_1$(side)=0.079   r$_2$(side)=0.177

P=31^d.3058   r$_1$(back)=0.079   r$_2$(back)=0.192

i=87°.15 Ω_1=12.750   V$_\gamma$= ———

T$_1$=8840 K Ω_2=2.093   F$_1$=1.00

T$_2$=4465 K   q=0.115   F$_2$=1.00

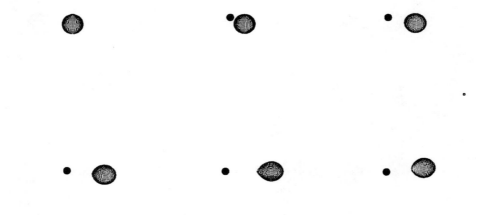

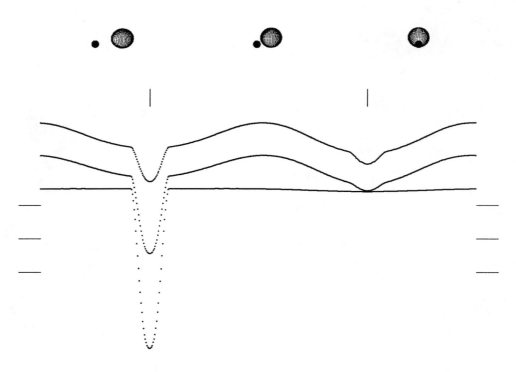

a=42.9 R$_\odot$	r$_1$(pole)=0.078	r$_2$(pole)=0.222
e= 0.000	r$_1$(point)=0.078	r$_2$(point)=0.327
ω= −−−	r$_1$(side)=0.078	r$_2$(side)=0.231
P=12^d.209	r$_1$(back)=0.078	r$_2$(back)=0.263
i=81°.0	Ω_1=13.008	V$_\gamma$=+33.5 km sec^{-1}
T$_1$=8620 K	Ω_2=2.156	F$_1$=1.00
T$_2$=4780 K	q=0.17	F$_2$=1.00

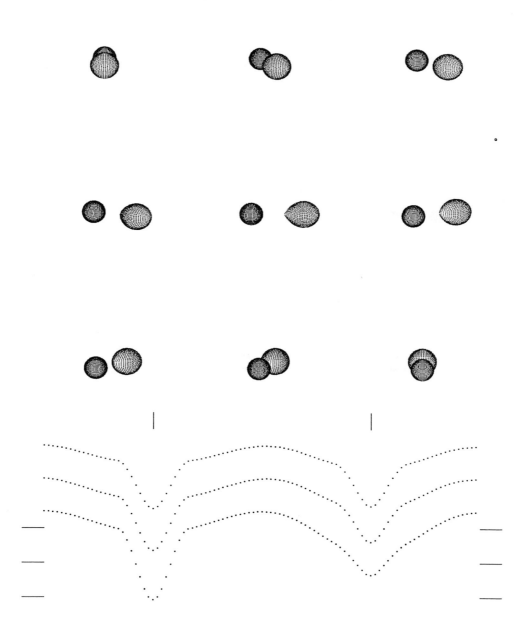

$a=48.83\ R_{\odot}$ $r_1(\text{pole})=0.218$ $r_2(\text{pole})=0.261$

$e=\ 0.000$ $r_1(\text{point})=0.221$ $r_2(\text{point})=0.379$

$\omega=\ ---$ $r_1(\text{side})=0.220$ $r_2(\text{side})=0.272$

$P=21^{\text{d}}.208$ $r_1(\text{back})=0.221$ $r_2(\text{back})=0.305$

$i=81°.0$ $\Omega_1=4.870$ $V_{\gamma}=+12.3\ \text{km sec}^{-1}$

$T_1=4786\ \text{K}$ $\Omega_2=2.466$ $F_1=1.00$

$T_2=4467\ \text{K}$ $q=0.3$ $F_2=1.00$

337

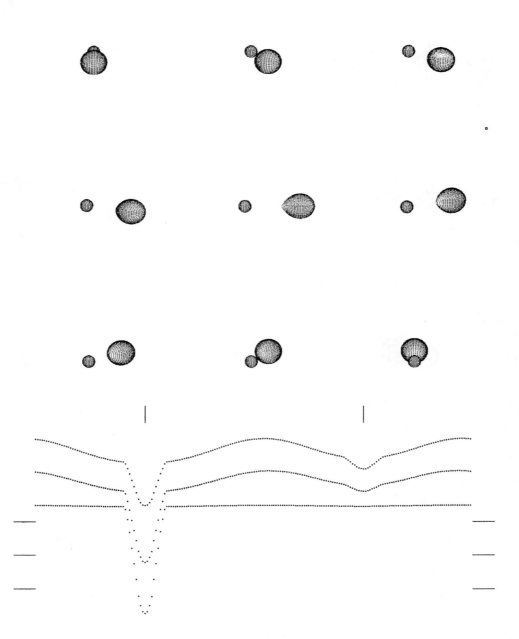

a=51.1 R$_\odot$ r$_1$(pole)=0.115   r$_2$(pole)=0.233

e= 0.000   r$_1$(point)=0.115   r$_2$(point)=0.341

ω= ———   r$_1$(side)=0.115   r$_2$(side)=0.242

P=11^{d}.113   r$_1$(back)=0.115   r$_2$(back)=0.275

i=78°.4 Ω_1=8.897   V$_\gamma$=−7.0 km sec^{-1}

T$_1$=15000 K Ω_2=2.233   F$_1$=1.00

T$_2$=6600 K   q=0.2   F$_2$=1.00

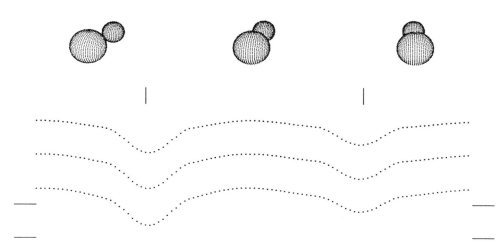

a=43.2 R$_\odot$	r$_1$(pole)=0.381	r$_2$(pole)=0.234
e= 0.040	r$_1$(point)=0.438	r$_2$(point)=0.264
ω= 118°.0	r$_1$(side)=0.400	r$_2$(side)=0.240
P=5^d.7325	r$_1$(back)=0.416	r$_2$(back)=0.254
i=68°.0	Ω$_1$=3.015	V$_\gamma$=+20.1 km sec^{-1}
T$_1$=31100 K	Ω$_2$=3.024	F$_1$=1.08
T$_2$=25000 K	q=0.4	F$_2$=1.08

a=47.75 R$_\odot$ r$_1$(pole)=0.037   r$_2$(pole)=0.061

e=0.189   r$_1$(point)=0.037   r$_2$(point)=0.061

ω=109°.8   r$_1$(side)=0.037   r$_2$(side)=0.061

P=24^d.5923   r$_1$(back)=0.037   r$_2$(back)=0.061

i=88°.45 Ω_1=28.371   V$_\gamma$=−1.8 km sec^{-1}

T$_1$=6310 K Ω_2=18.092   F$_1$=1.49

T$_2$=5161 K   q=1.034   F$_2$=1.49

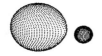
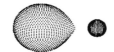

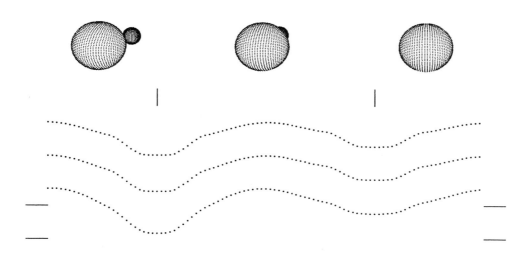

a= 51.86 $R_\odot$	r_1(pole)=0.453	r_2(pole)=0.161
e= 0.000	r_1(point)=0.618	r_2(point)=0.166
ω= ---	r_1(side)=0.487	r_2(side)=0.162
P=7^d.8482	r_1(back)=0.512	r_2(back)=0.165
i=76°.9	Ω_1=2.488	V_γ=−40.5 km sec^{-1}
T_1=29000 K	Ω_2=3.262	F_1=1.00
T_2=17056 K	q=0.31	F_2=1.00

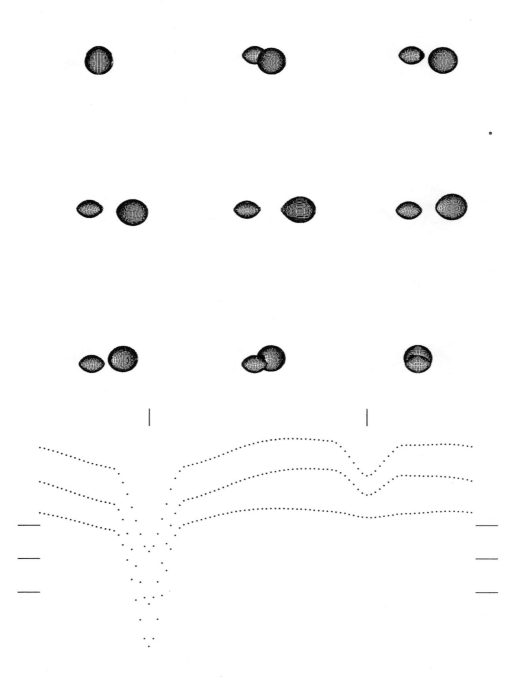

a= 49.5 R$_\odot$ r$_1$(pole)=0.173   r$_2$(pole)=0.256

e= 0.000   r$_1$(point)=0.259   r$_2$(point)=0.372

ω= ---   r$_1$(side)=0.241   r$_2$(side)=0.266

P=15^د.1907   r$_1$(back)=0.249   r$_2$(back)=0.299

i=83°.10 Ω_1=6.066   V$_\gamma$=−13.4 km sec^{-1}

T$_1$=22700 K Ω_2=2.416   F$_1$=6.66

T$_2$=7500 K   q=0.277   F$_2$=1.0

a= 53.1 $R_\odot$	r_1(pole)= ---	r_2(pole)=0.574
e= 0.092	r_1(point)= ---	r_2(point)=0.672
ω=154°.0	r_1(side)= ---	r_2(side)=0.590
P=8^d.9649	r_1(back)= ---	r_2(back)=0.604
i=80°.0	Ω_1= ---	V_γ=−7.2 km sec^{-1}
T_1= ---	Ω_2=17.559	F_1= ---
T_2=22000 K	q=13.0	F_2=0.50

a= 43.03 R$_\odot$	r$_1$(pole)= ---	r$_2$(pole)=0.418
e= 0.000	r$_1$(point)= ---	r$_2$(point)=0.426
ω= ---	r$_1$(side)= ---	r$_2$(side)=0.422
P=3^d.4112	r$_1$(back)= ---	r$_2$(back)=0.424
i=87°.0	Ω_1= ---	V$_\gamma$=−67 km sec^{-1}
T$_1$= ---	Ω_2=23. 372	F$_1$= ---
T$_2$=45000 K	q=21.0	F$_2$=0.50

Notes

TW And: Popper (PASP **94**, 945) and Hiltner (ApJ **109**, 95) have identical values for K_1. Popper also gives K_2. The resulting mass ratio is the same as the photometric mass ratio.

KO Aql: The mass ratio taken from the photometric reference is neither q_{ptm} nor q_{rv} (The binary is detached in the photometric solution and there are no velocities for the secondary.).

RY Aqr: The published q_{rv} by Popper is 0.213. Helt adopted 0.201 by private communication from Popper.

DV Aqr: The secondary temperature was estimated from the value of M_V given for the secondary. The mass ratio should be taken with caution, as this system is well-detached and single-lined.

IU Aur: We used the solution with $i=88°.2$, corresponding to epoch 1984.4. The system has third light.

LY Aur: The secondary temperature should be lower than the primary temperature. The value of r_2(side) for $q=0.375$ is 0.289, not 0.366 as given in the photometric reference.

SS Cam: The mass ratio is q_{rv}.

AT Cam: We used solution 2 in the photometric solution, which seems more plausible in terms of evolutionary theory.

AW Cam: The photometric reference mentions the fact that the mass of the primary is approximately $17M_\odot$ if the mass function value of 0.113 is used. This is too high for an A0 V star.

AZ Cam: The evolutionary configuration of the system seems impossible. A double-lined radial velocity solution is badly needed to determine the masses.

UW CMa: There are serious problems with all of the published light curve and radial velocity solutions. A reasonably good velocity curve for the secondary is badly needed.

CW CMa: The mass ratio is q_{rv}. The temperatures were obtained from the spectral types and not from the solution in the photometric reference.

VZ CVn: The mass ratio is a guess based on the spectral types.

X Car: The masses are anomalously low for the spectral type.

V348 Car: The elements of this system should be taken with some caution as the depths of the eclipses are not very well defined.

YZ Cas: The mass ratio is q_{rv}.

AO Cas: The value of q_{ptm} is unlikely to be meaningful. See *Observatory* **108**, no. 1086, page 174 for a discussion of widely varying mass ratios.

PV Cas: The value of ω was computed from Popper's formula (p. 672 of the radial velocity reference) for 2000 A.D.

V364 Cas: The photometric reference does not list the mass ratio or the temperatures. These were estimated from the spectral types.

BH Cen: The absolute dimensions given in the photometric reference are based on the membership of the system in IC 2944.

XY Cep: The mass ratio is that needed to produce the side radius of the secondary star, assuming that it fills its lobe.

BE Cep: The temperature difference is surprisingly large for the large degree of overcontact.

XY Cet: The mass ratio is q_{rv}.

RS Cha: Andersen discusses the somewhat unusual evolutionary configuration in the radial velocity reference.

U CrB: We adopted the mass ratio from the radial velocity solution.

RV Crv: The authors of the photometric solution used a value of 3.2 for the bolometric albedo of the secondary, and we have done so here. That is why the UV light curve shows such strong reflection.

MR Cyg: This system has received considerable attention. We have adopted the $a_1 \sin i$ by Hill and Hutchings and the mass ratio from Linnell and Kallrath. This procedure results in masses which are too low theoretically, but they are measured (although weakly) values.

MY Cyg: The spectroscopic reference labels the cooler star as the primary, which differs from the photometric reference and our adopted labelling convention. The primary eclipse is the one of the more massive, hotter, and slightly larger star.

V382 Cyg: The photometric solution comes from the application of an ellipsoid model to an overcontact system.

| V470 Cyg: | The values of q_{ptm} (1.14) and q_{rv} (0.88) are quite different. Since the inclination is around 50° and the system is detached, the q_{ptm} is certainly unreliable. We used q_{rv} and iterated to match the radii in the photometric solution. |

V470 Cyg: The values of q_{ptm} (1.14) and q_{rv} (0.88) are quite different. Since the inclination is around 50° and the system is detached, the q_{ptm} is certainly unreliable. We used q_{rv} and iterated to match the radii in the photometric solution.

V1425 Cyg: We used Lee's solution I. Solution II is for star 1 filling the lobe, which is astrophysically implausible.

V1727 Cyg: This system is an X-ray binary.

RZ Dra: The primary star would be about 0.6 $M_\odot$ at 9070 K. New velocities are needed.

CW Eri: The mass ratio is q_{rv}.

RX Gem: The solution takes into account the ring of material around the primary.

u Her: Eccentricity of order 0.05 is not supported by the ephemeris (see A&Ap **66**,161).

UX Her: T_1 seems anomalously low for the estimated M_1 of about 3.3 $M_\odot$. New spectroscopic work would be useful.

Y Hyi: The temperatures were obtained from the spectral types.

RT Lac: This is a very difficult binary to understand. The deeper eclipse is that of the spectroscopically cooler star. Probably RT Lac has some important feature which has not yet been modeled.

UV Leo: The photometric solution is at odds with previous Russell-Merrill solutions and with evolutionary ideas.

RR Lyn: The mass ratio is q_{rv}. The temperatures were obtained from the spectral types.

FL Lyr: We used q_{rv} and matched the radii from the photometric solution for this detached binary.

RU Mon: Struve had essentially no confidence in his velocity amplitude for this system, because the lines are badly blended. Using his $a \sin i$ gives ridiculous masses. We used a from Khaliullina *et al.* (MNRAS **216**, 909) which is not based on velocities of the eclipsing pair, but on the third body orbit and the mass-luminosity law. Their value of a is 18.8 $R_\odot$ while Struve's numbers give 6.22 $R_\odot$.

VV Mon: The mass ratio is q_{rv}.

AR Mon: The difference in primary and secondary eclipse depths requires a larger temperature difference than that found by Popper. The secondary star radii given here are a little smaller than those given by Popper and correspond to exact lobe filling for q=0.30.

δ Ori: A companion 0".15 away contributes a substantial amount of third light, which we did not include in the light curves.

ST Per: We adjusted the mass ratio until the value of r_2(side) matched the value given in the photometric reference. If q is kept fixed at 0.13 as given in the reference, then r_2(side)=0.213 instead of 0.230.

DM Per: The mass ratio is q_{rv}.

IW Per: Temperatures were obtained from spectral types.

IZ Per: The mass ratio had to be adjusted to agree with the relative side radius in the photometric solution. If q=0.25 (as given in the solution), r_2(side)=0.258 instead of 0.290.

LX Per: The mass ratio is q_{rv}.

β Per: The mass ratio is q_{ptm}.

AE Phe: We adopted star 2 (*i.e.*, the star eclipsed at phase 0.5) as the lower mass star. According to the authors this system is of W-type. However there is improved solution convergence if it is treated as an A-type. This is done by introducing a phase shift of 0.5 and q<1.

SZ Psc: According to the photometric solution, the secondary has cool spots. However, we were unable to translate the spot parameters into those used by the WD program.

KX Pup: The mass ratio in the photometric reference is an assumption.

NO Pup: The mass ratio was an assumption by the authors and is neither photometric nor spectroscopic. Third light is present in the system.

VV Pyx: The mass ratio is q_{rv}.

V906 Sco: We computed absolute dimensions by using Abt's value of K_1 (since the primary velocity curve looks much better than the secondary) and q_{ptm}.

CD Tau: The mass ratio is q_{rv}.

HU Tau: The mass ratio in the photometric reference is an assumption.

λ Tau: The elements from the photometric solution require this system to be detached, and there is no strong evidence for its being semi-detached.

TX UMa: T_1=13,500 K from the Hill and Hutchings solution seems a little high for a B8 V star. Perhaps 12,000 K would be better.

DN UMa: The mass ratio is q_{rv}. The photometric solution contains a large amount of third light.

W UMi: We adopted the side radius of the secondary from the photometric solution and found the mass ratio by assuming the system to be semi-detached. It is possible that the system is detached.

RT UMi: The photometric solution is for a slightly detached configuration. The authors' statement that they used Mode 5 may be a misprint.

RU UMi: The solution in the photometric reference is for a very slightly detached configuration, but it is practically semi-detached.

BH Vir: Abt's mass ratio presents a problem: the more massive star is less evolved in a detached system. However, the radial velocity curve does have considerable scatter.

DL Vir: We adjusted the mass ratio until the value of $r_2(side)$ matched the value in the photometric reference.

Z Vul: The mass ratio of 0.43 is based on $r_2(side)$ under the semi-detached assumption. The value of q_{rv} is 0.41.

RS Vul: The relative dimensions of the secondary given in the photometric solution are not consistent with the mass ratio and the assumption of a semi-detached configuration. We used Mode 2 and matched the side radius given in the photometric solution. The system may be semi-detached. The two radial velocity solutions are practically identical.

HD 27130: The absolute dimensions are based on membership in the Hyades cluster.

HD 77581: The parameter values are our best estimates from the extensive literature on this X-ray binary. It was impossible to satisfy all observational constraints with the optical star rotating synchronously, as pointed out in ApJ **234**,1054.

HD 149779: The mass ratio in the photometric reference is an assumption.

HD 153919: The parameter values are our best estimates from the literature on this system. All observational constraints could not be satisfied with the optical star rotating synchronously.

References

References for the photometric (P) and spectroscopic (S) data are listed below by constellation. Abbreviations in the references are as follows:

AA	Acta Astronomica
AJ	Astronomical Journal
ApJ	Astrophysical Journal
ApJSup	Astrophysical Journal Supplement Series
ApSpSc	Astrophysics and Space Science
A&Ap	Astronomy and Astrophysics
A&ApSup	Astronomy and Astrophysics Supplement Series
BAC	Bulletin of the Astronomical Institutes of Czechoslovakia
ChA&Ap	Chinese Astronomy and Astrophysics
IBVS	Information Bulletin of Variable Stars
MNRAS	Monthly Notices of the Royal Astronomical Society
MSAI	Memorie della Societe Astronomica Italiana
PASJ	Publications of the Astronomical Society of Japan
PASP	Publications of the Astronomical Society of the Pacific
PDAO	Publications of the Dominion Astrophysical Observatory
PRAO	Publications of the Rothney Astrophysical Observatory

Andromeda

RT   P:A&Ap **103**,57   S:ApJ **103**,291   Mode 2

TW   P:A&Ap **24**,131   S:ApJ **109**,95 and PASP **94**,945   Mode 5

AB   P:ApJ **335**,319   S:same   Mode 3

AD   P:A&ApSup **45**,499   S:none   Mode 2

AN   P:A&Ap **114**,74   S:Trieste Contr. No. 287 (DAO)   Mode 2

BL   P:AA **35**,327   S:none   Mode 3

CN   P:PASP **97**,310   S:none   Mode 3

DS   P:AJ **95**,1466   S:same   Mode 2

Antlia

S   P:A&ApSup **47**,211   S:ApJ **124**,208   Mode 3

Aquarius

RY   P:A&Ap **172**,155   S:PASP **94**,945   Mode 5

ST   P:AA **35**,327 S:none Mode 3

BW   P:ApSpSc **120**,9 S:A&ApSup **69**,397 Mode 2

CX   P:MNRAS **223**,607 S:same Mode 5

DV   P:PASP **97**,62 S:A&ApSup **24**,29 Mode 2

EE   P:MNRAS **232**,147 S:same Mode 2

Aquila

KO   P:A&ApSup **45**,85 S:ApJ **76**,544 Mode 2

OO   P:ApJ **340**,458 S:same Mode 3

QY   P:ApSpSc **76**,111 S:ApJ **103**,76 Mode 2

V337 P:A&ApSup **45**,499 S:none Mode 5

V346 P:A&ApSup **45**,85 S:none Mode 5

V805 P:A&ApSup **32**,351 S:ApJ **244**,541 Mode 2

V889 P:A&Ap **115**,321 S:none Mode 2

V1182 P:MNRAS **250**,209 S:same and MNRAS **225**,961 Mode 2

Ara

V535 P:ApJ **222**,917 S:A&ApSup **36**,287 (DAO) Mode 3

V539 P:MSAI **50**,571 S:A&Ap **118**,255 Mode 2

Aries

RX   P:A&ApSup **42**,195 S:none Mode 2

SS   P:AA **34**,445 S:none Mode 3

Auriga

SX   P:MNRAS **224**,649 S:same Mode 2

TT   P:MNRAS **250**,209 S:same and A&Ap **162**,62 Mode 5

WW   P:MSAI **50**,574   S:Ann. Tokyo Astr. Obs. 2nd series **16**,22 (DAO)   Mode 2

ZZ   P:AJ **90**,115   S:none   Mode 2

AR   P:MSAI **50**,575   S:PASP **48**,24   Mode 2

EO   P:ApSpSc **70**,461   S:Pub. Amer. Ast. Soc. **10**,332 (DAO)   Mode 2

HP   P:AA **33**,159   S:none   Mode 2

HS   P:AJ **91**,383   S:same   Mode 2

IM   P:A&ApSup **60**,389   S:same   Mode 2

IU   P:A&Ap **183**,161   S:A&Ap **59**,9   Mode 5

LY   P:A&Ap **62**,291   S:A&Ap **31**,1   Mode 2

Boötes

SS   P:AJ **88**,1257   S:AJ **100**,247   Mode 2

TY   P:AJ **98**,2287   S:PRAO #56   Mode 3

XY   P:AJ **82**,648   S:none   Mode 1

ZZ   P:A&ApSup **32**,347   S:AJ **88**,1242   Mode 2

44i   P:A&ApSup **45**,187   S:ApJ **97**,394   Mode 3

Camelopardalis

Y   P:MSAI **50**,591   S:ApJ **111**,658   Mode 5

SS   P:AA **29**,243   S:AJ **100**,247   Mode 2

SV   P:MSAI **50**,592   S:ApJ **118**,262   Mode 2

SZ   P:ApSpSc **76**,23   S:BAC **31**,321 (DAO)   Mode 2

TU   P:A&ApSup **42**,15   S:AJ **76**,544   Mode 2

AO   P:ApSpSc **113**,25   S:none   Mode 3

AT   P:A&Ap **141**,266   S:none   Mode 2

AW   P:A&ApSup **52**,311   S:MSAI **38**,509   Mode 5

AZ   P:A&Ap **141**,266   S:none   Mode 2

Cancer

S   P:W. Van Hamme and R.E. Wilson, preprint   S:same   Mode 5

RZ   P:AJ **98**,1002   S:ApJ **208**,142   Mode 2

TX   P:A&Ap **48**,349   S:ApJ **133**,133   Mode 3

AD   P:AJ **98**,2287   S:none   Mode 3

AH   P:A&ApSup **58**,405   S:MNRAS **186**,729   Mode 3

Canis Major

R   P:MSAI **50**,602   S:ApJ **297**,250   Mode 5

UW   P:ApJ **222**,924   S:*Observatory* **109**,74   Mode 3

CW   P:MSAI **50**,609   S:ApJ **261**,612   Mode 2

Canis Minor

XZ   P:PASP **102**,646   S:none   Mode 5

Canes Venatici

VZ   P:A&Ap **56**,75   S:none   Mode 2

Capricornus

δ   P:AJ **97**,499   S:PDAO **11**,395 (DAO)   Mode 2

Carina

X   P:ApSpSc **99**,191   S:ApJ **115**,134   Mode 2

ST   P:MNRAS **206**,305   S:none   Mode 2

GW   P:ApSpSc **99**,191   S:none   Mode 5

QX   P:A&Ap **121**,271   S:A&Ap **121**,271   Mode 2

V348　P:MNRAS **213**,75　S:same　Mode 3

Cassiopeia

TV　　P:AA **34**,46　S:PDAO **2**,141 (DAO)　Mode 2

TX　　P:MNRAS **216**,663　S:none　Mode 2

YZ　　P:MSAI **50**,616　S:ApJ **251**,591　Mode 2

AE　　P:AA **34**,281　S:none　Mode 2

AO　　P:ApJ **223**,202　S:ApJSup **4**,157　Mode 1

IT　　P:ApSpSc **155**,53　S:none　Mode 2

MN　　P:AJ **82**,290　S:none　Mode 2

PV　　P:AJ **93**,672　S:same　Mode 2

V364　P:ApSpSc **106**,273　S:none　Mode 2

V523　P:A&Ap **170**,43　S:AJ **90**,354　Mode 3

Centaurus

RR　　P:MNRAS **189**,907　S:MNRAS **209**,645　Mode 3

ST　　P:AJ **86**,1546　S:none　Mode 2

SV　　P:MNRAS **176**,625　S:A&Ap **110**,246　Mode 3

SZ　　P:MSAI **50**,627　S:A&Ap **45**,203　Mode 2

BH　　P:AJ **89**,872　S:none　Mode 3

KT　　P:A&ApSup **22**,263　S:none　Mode 2

V346　P:A&Ap **160**,310　S:PASP **90**,728　Mode 2

V757　P:A&ApSup **58**,405　S:PASP **94**,195　Mode 3

Cepheus

U　　　P:MNRAS **187**,699　S:ApJ **244**,546　Mode 5

VW   P:A&Ap **218**,141   S:same   Mode 6

WX   P:A&ApSup **32**,351   S:AJ **93**,672   Mode 2

WZ   P:AA **36**,105   S:none   Mode 3

XX   P:A&Ap **156**,38   S:ApJ **103**,76   Mode 2

XY   P:ApSpSc **66**,143   S:AJ **76**,544   Mode 5

ZZ   P:A&ApSup **33**,91   S:ApJ **106**,112   Mode 2

AH   P:A&Ap **221**,49   S:MNRAS **223**,513   Mode 2

BE   P:PASP **98**,662   S:none   Mode 3

CQ   P:ApJ **265**,961   S:same   Mode 3

CW   P:MNRAS **250**,209   S:same and ApJ **188**,559   Mode 2

EG   P:AA **34**,433   S:none   Mode 2

EI   P:ApSpSc **32**,285   S:ApJ **166**,361   Mode 2

EK   P:MSAI **50**,641   S:ApJ **271**,717   Mode 2

ER   P:MNRAS **184**,33   S:none   Mode 3

GK   P:ApJ **179**,539   S:Asiago Contr. #168 (DAO)   Mode 2

GT   P:A&ApSup **55**,403   S:Pub. David Dunlap Obs. **2**,417 (DAO)   Mode 5

GW   P:AA **34**,217   S:none   Mode 3

NN   P:A&ApSup **51**,27   S:none   Mode 2

Cetus

TV   P:MSAI **50**,644   S:ApJ **154**,198   Mode 2

TW   P:A&ApSup **47**,211   S:ApJ **111**,658   Mode 3

TX   P:A&Ap **70**,355   S:none   Mode 2

VY   P:A&Ap **161**,264   S:none   Mode 3

XY   P:ApSpSc **38**,79   S:ApJ **169**,549   Mode 2

Chamaeleon

RS   P:A&Ap **85**,259   S:A&Ap **44**,445   Mode 2

RZ   P:A&Ap **85**,259   S:A&Ap **44**,349   Mode 2

Coma Berenices

RW   P:ApJ **319**,325   S:AJ **90**,109   Mode 3

RZ   P:MSAI **50**,647   S:MNRAS **203**,1   Mode 3

Corona Australis

TZ   P:A&ApSup **42**,195   S:none   Mode 2

Corona Borealis

U   P:A&Ap **61**,469   S:PDAO **15**,419 (DAO)   Mode 5

RT   P:AJ **100**,247   S:same   Mode 2

RW   P:A&ApSup **42**,195   S:ApJ **79**,89   Mode 2

α   P:ApSpSc **26**,371   S:ApJ **91**,1428   Mode 2

Corvus

W   P:ApJ **231**,502   S:none   Mode 3

RV   P:MNRAS **223**,595   S:same   Mode 2

Cygnus

Y   P:A&ApSup **39**,255   S:Pub. Crimean Ap. Obs. **43**,71 (DAO)   Mode 2

SW   P:AJ **96**,747   S:ApJ **104**,253   Mode 6

UZ   P:ApSpSc **82**,189   S:none   Mode 5

VW   P:ApSpSc **82**,189   S:ApJ **103**,76   Mode 5

WW   P:A&ApSup **39**,265   S:ApJ **104**,253   Mode 5

BR   P:A&Ap **96**,409   S:none   Mode 5

CG   P:IBVS 3305,3398   S:AJ **90**,761   Mode 2

MR   P:ApJ **166**,605   S:A&Ap **23**,357   Mode 2

MY   P:A&ApSup **39**,255   S:ApJ **169**,549   Mode 2

V382 P:A&ApSup **33**,91   S:PASP **64**,219   Mode 3

V388 P:MNRAS **203**,235   S:none   Mode 4

V442 P:AJ **94**,712   S:same   Mode 2

V453 P:A&ApSup **33**,91   S:same   Mode 2

V463 P:BAC **33**,187   S:none   Mode 5

V470 P:A&Ap **109**,368   S:PASP **58**,247   Mode 2

V478 P:A&ApSup **53**,363   S:AJ **101**,600   Mode 2

V548 P:AA **33**,163   S:Pub. David Dunlap Obs. **2**,255 (DAO)   Mode 5

V1073 P:ApJ **222**,917   S:Pub. David Dunlap Obs. **2**,417 (DAO)   Mode 3

V1143 P:A&Ap **141**,1   S:same   Mode 2

V1425 P:ApJ **338**,1016   S:none   Mode 2

V1727 P:ApJ **243**,900   S:same   Mode 4

Delphinus

W   P:MSAI **50**,672   S:ApJ **104**,253   Mode 5

DM   P:A&ApSup **67**,87   S:none   Mode 2

Draco

RZ   P:ChA&Ap **6**,199   S:ApJ **103**,76   Mode 5

UZ   P:AJ **97**,822   S:same   Mode 2

AI   P:MSAI **50**,675   S:AA **28**,41   Mode 5

AX   P:ChA&Ap **13**,216   S:none   Mode 2

BH   P:A&ApSup **45**,499   S:Abh. Hamburger Sternw. Bd.8, No. 5 (DAO)   Mode 2

BS   P:ApSpSc **79**,359   S:ApJ **166**,361   Mode 2

BV   P:AJ **92**,666   S:same   Mode 3

BW   P:AJ **92**,666   S:same   Mode 3

CM   P:ApJ **218**,444   S:same   Mode 2

Equuleus

S   P:A&ApSup **36**,273   S:BAC **17**,295   Mode 5

Eridanus

RU   P:ApSpSc **81**,209   S:PASJ **36**,277   Mode 5

WX   P:A&ApSup **52**,311   S:same   Mode 5

YY   P:A&ApSup **49**,123   S:A&Ap **159**,142   Mode 3

AS   P:A&Ap **141**,1   S:same   Mode 5

BW   P:ApSpSc **88**,197   S:none   Mode 2

CW   P:MSAI **50**,682   S:AJ **88**,1242   Mode 2

Gemini

RW   P:A&ApSup **36**,273   S:AJ **76**,544   Mode 5

RX   P:A&Ap **38**,225   S:ApJ **104**,376   Mode 5

RY   P:AJ **100**,1981   S:same   Mode 5

YY   P:AJ **83**,618   S:ApJ **193**,389   Mode 2

AF   P:PASP **94**,926   S:none   Mode 5

GW   P:A&ApSup **46**,185   S:none   Mode 5

Grus

W   P:AA **37**,41   S:A&Ap **32**,429   Mode 2

Hercules

    u   P:MNRAS **211**,943   S:same   Mode 5

    RX   P:A&ApSup **33**,91   S:ApJ **129**,659   Mode 2

    SZ   P:A&ApSup **45**,85   S:none   Mode 5

    TT   P:A&Ap **126**,94   S:ApJ **86**,153   Mode 2

    TX   P:A&ApSup **32**,351   S:ApJ **162**,925   Mode 2

    UX   P:A&ApSup **40**,57   S:ApJ **86**,153   Mode 5

    AD   P:A&ApSup **39**,235   S:PASP **90**,312   Mode 5

    AK   P:ApSpSc **52**,387   S:ApJ **79**,89   Mode 3

    HS   P:ApSpSc **76**,111   S:ApJ **101**,114   Mode 2

    LT   P:A&ApSup **52**,311   S:none   Mode 5

    V338 P:A&ApSup **39**,255   S:none   Mode 5

    V624 P:MSAI **50**,695   S:AJ **89**,1057   Mode 2

Horologium

    SY   P:A&Ap **161**,264   S:none   Mode 3

Hydra

    TT   P:W. Van Hamme and R.E. Wilson, preprint   S:same   Mode 5

    VZ   P:MSAI **50**,697   S:AJ **141**,126   Mode 2

    WY   P:A&Ap **103**,349   S:same   Mode 2

    AI   P:A&Ap **66**,377   S:AJ **95**,190   Mode 2

    HS   P:A&Ap **42**,303   S:ApJ **166**,361   Mode 2

    KM   P:A&Ap **130**,102   S:same   Mode 2

χ^2 P:A&Ap **67**,15   S:A&Ap **44**,445   Mode 2

Hydrus

Y   P:ApSpSc **119**,345   S:none   Mode 5

Indus

RS   P:MNRAS **232**,147   S:same   Mode 2

RY   P:MNRAS **206**,305   S:none   Mode 2

Lacerta

RT   P:ApJ **227**,907   S:AJ **91**,583   Mode 2

SW   P:PASP **96**,634   S:ApJ **109**,436   Mode 3

VY   P:AA **34**,207   S:none   Mode 5

AR   P:AJ **97**,848   S:ApJ **113**,299   Mode 2

AW   P:AJ **88**,1679   S:none   Mode 3

CM   P:A&ApSup **32**,351   S:ApJ **154**,191   Mode 2

CO   P:ApSpSc **89**,5   S:AA **17**,245   Mode 2

EM   P:A&ApSup **51**,435   S:none   Mode 3

Leo

UV   P:MSAI **50**,706   S:ApJ **141**,126   Mode 2

XY   P:MSAI **50**,707   S:ApJ **317**,333   Mode 3

AM   P:PASP **96**,646   S:none   Mode 3

Leo Minor

T   P:A&ApSup **36**,273   S:ApJ **104**,253   Mode 5

Libra

ES   P:ApSpSc **153**,273   S:same   Mode 2

δ　　　P:A&ApSup **37**,513　　S:ApJ **221**,608　　Mode 5

Lupus

　　FT　　P:AA **36**,113　　S:MNRAS **208**,135　　Mode 5

Lynx

　　RR　　P:ApSpSc **30**,433　　S:ApJ **169**,549　　Mode 2

　　SW　　P:ApJ **234**,1054　　S:BAC **28**,120　　Mode 5

　　UU　　P:PASJ **35**,131　　S:same　　Mode 2

　　UV　　P:PASP **94**,350　　S:none　　Mode 3

Lyra

　　TZ　　P:AA **35**,327　　S:none　　Mode 5

　　FL　　P:MSAI **50**,711　　S:AJ **91**,383　　Mode 2

　　V361　P:AJ **99**,1207　　S:none　　Mode 4

Mensa

　　TZ　　P:A&Ap **175**,60　　S:same　　Mode 2

　　UX　　P:J.Andersen, J.V. Clausen, & P. Magain, Harvard-Smithsonian Center for Astrophysics - Preprint Series #2734　　S:same　　Mode 2

Monoceros

　　RU　　P:MNRAS **216**,909　　S:ApJ **102**,74 and MNRAS **216**,909　　Mode 2

　　RW　　P:AJ **100**,1981　　S:same　　Mode 5

　　TU　　P:MSAI **50**,714　　S:ApJ **102**,433　　Mode 5

　　VV　　P:A&ApSup **35**,291　　S:AJ **96**,1040　　Mode 2

　　AO　　P:MSAI **50**,715　　S:ApJ **102**,74　　Mode 2

　　AR　　P:A&ApSup **39**,255　　S:ApJ **208**,142　　Mode 5

　　AU　　P:MNRAS **199**,131　　S:PASP **94**,113　　Mode 5

IM   P:A&ApSup **33**,91   S:PDAO **6**,70 (DAO)   Mode 2

Musca

TU   P:ApSpSc **76**,23   S:A&Ap **45**,107   Mode 3

Octans

UZ   P:A&Ap **130**,97   S:none   Mode 3

Ophiuchus

U   P:A&ApSup **33**,91   S:AJ **101**,600   Mode 2

RV   P:A&ApSup **39**,273   S:none   Mode 5

WZ   P:A&ApSup **32**,351   S:ApJ **141**,126   Mode 2

V451   P:A&Ap **167**,287   S:ApJ **166**,361   Mode 2

V502   P:A&ApSup **49**,123   S:ApJ **130**,789   Mode 3

V508   P:A&Ap **231**,365   S:same   Mode 3

V566   P:A&Ap **152**,25   S:same   Mode 3

V1010   P:AJ **101**,1828   S:PASP **89**,74   Mode 4

Orion

VV   P:ApJ **197**,379   S:A&ApSup **22**,19   Mode 2

BM   P:ApSpSc **103**,115   S:ApJ **205**,462   Mode 5

ER   P:A&ApSup **47**,211   S:PASP **56**,34   Mode 3

δ P:ApJ **248**,249   S:Pub. Mich. Obs. **1**,118 (DAO)   Mode 2

Pegasus

U   P:ChA&Ap **12**,223   S:same   Mode 3

AQ   P:AJ **96**,747   S:ApJ **103**,76   Mode 6

AW   P:AJ **96**,747   S:same   Mode 5

BB   P:AJ **90**,515   S:none   Mode 3

BK   P:AJ **86**,102   S:AJ **88**,1242   Mode 2

BO   P:PASP **98**,1325   S:none   Mode 2

BX   P:AJ **90**,515   S:none   Mode 3

DI   P:A&ApSup **42**,195   S:none   Mode 2

EE   P:MSAI **50**,730   S:ApJ **281**,268   Mode 2

Perseus

RW   P:AJ **95**,1828   S:ApJ **102**,74 and D.M. Popper, priv. comm.   Mode 2

RY   P:AJ **92**,1168   S:AJ **92**,1168   Mode 5

ST   P:MSAI **50**,732   S:AJ **76**,544   Mode 5

AG   P:MNRAS **250**,209   S:same and ApJ **188**,559   Mode 2

DM   P:MNRAS **222**,167   S:same   Mode 5

IQ   P:MSAI **50**,734   S:ApJ **295**,569   Mode 2

IW   P:ApSpSc **68**,355   S:PDAO **6**,214   Mode 2

IZ   P:MSAI **50**,735   S:A&Ap **2**,388 (DAO)   Mode 2

LX   P:ApSpSc **112**,273   S:AJ **96**,1040   Mode 2

β   P:ApJ **342**,1061   S:ApJ **168**,443   Mode 5

Phoenix

AE   P:A&Ap **152**,25   S:same   Mode 3

AG   P:PASP **95**,347   S:none   Mode 2

AI   P:sub. to ApJ, E.F.Milone, C.R. Stagg, R.L. Kurucz   S:same   Mode 0

ζ   P:A&Ap **46**,205   S:A&Ap **118**,255   Mode 2

Pisces

    Y   P:A&ApSup **39**,265   S:ApJ **104**,253   Mode 5

    SZ   P:ApJ **227**,907   S:AJ **81**,250   Mode 2

    VZ   P:AA **234**,177   S:AJ **97**,532   Mode 5

Piscis Austrinus

    RW   P:ApJ **231**,502   S:none   Mode 3

Puppis

    V   P:A&Ap **128**,17   S:same   Mode 5

    UZ   P:ApSpSc **153**,269   S:ApJ **102**,74   Mode 3

    XZ   P:MSAI **50**,746   S:AJ **76**,544   Mode 5

    AU   P:ApJ **222**,917   S:none   Mode 3

    KX   P:A&ApSup **22**,263   S:none   Mode 2

    NO   P:A&Ap **50**,79   S:none   Mode 2

Pyxis

    RZ   P:MNRAS **227**,481   S:same   Mode 3

    TY   P:A&Ap **101**,7   S:A&Ap **39**,131   Mode 2

    VV   P:A&Ap **134**,147   S:same   Mode 2

Sagitta

    U   P:AJ **92**,1168   S:same   Mode 5

Sagittarius

    XZ   P:IBVS 1827   S:ApJ **109**,439   Mode 2

    V505 P:A&ApSup **39**,273   S:ApJ **109**,100   Mode 5

    V1647 P:A&Ap **58**,121   S:A&Ap **145**,206   Mode 2

Scorpius

    μ^1 P:AJ **84**,236   S: *Festschrift für Ellis Strömgren*, 258 (DAO)   Mode 5

    V499 P:ApSpSc **76**,23   S:none   Mode 2

    V701 P:A&Ap **61**,137   S:MNRAS **226**,899   Mode 3

    V760 P:A&Ap **151**,329   S:same   Mode 2

    V861 P:MNRAS **203**,1021   S:*Observatory* vol. III #1100, p. 23   Mode 4

    V906 P:ApJ **201**,792   S:ApJ **159**,919   Mode 2

Sculptor

    RT   P:ApSpSc **100**,117   S:MNRAS **223**,581   Mode 5

Scutum

    RS   P:ApSpSc **99**,191   S:none   Mode 2

    RZ   P:ApJ **289**,748   S:same   Mode 6

Serpens

    AU   P:AA **36**,113   S:none   Mode 3

Taurus

    RW   P:AJ **100**,1981   S:ApJ **110**,438   Mode 5

    RZ   P:ApJ **182**,539   S:ApJ **111**,658   Mode 1

    CD   P:AJ **81**,855   S:ApJ **166**,361   Mode 2

    HU   P:A&Ap **97**,410   S:MSAI **38**,459   Mode 2

    λ P:A&Ap **62**,291   S:ApJ **263**,289   Mode 5

Telescopium

    HO   P:A&ApSup **45**,499   S:none   Mode 2

Triangulum

X   P:A&ApSup **39**,265   S:ApJ **104**,253   Mode 5

Triangulum Australe

RR   P:A&ApSup **45**,85   S:none   Mode 5

Tucana

AQ   P:MNRAS **223**,581   S:same   Mode 3

CF   P:ApSpSc **133**,45   S:MNRAS **197**,769   Mode 2

Ursa Major

TX   P:MSAI **50**,770   S:PASP **80**,192   Mode 5

TY   P:A&ApSup **51**,97   S:none   Mode 3

VV   P:MSAI **157**,773   S:ApJ **112**,184   Mode 5

AW   P:ApJ **260**,744   S:MNRAS **195**,931   Mode 3

DN   P:ApSpSc **125**,181   S:PASP **98**,1312   Mode 2

Ursa Minor

W   P:A&ApSup **40**,57   S:ApJ **102**,470   Mode 5

RT   P:A&Ap **96**,328   S:none   Mode 2

RU   P:AA **35**,327   S:same   Mode 2

Vela

S   P:A&ApSup **36**,273   S:ApJ **116**,35   Mode 5

CV   P:MSAI **50**,781   S:A&Ap **44**,355   Mode 2

EO   P:A&ApSup **22**,263   S:none   Mode 2

Virgo

AG   P:AA **36**,121   S:PASP **84**,382   Mode 3

AH   P:PASP **96**,646   S:ApJ **107**,96   Mode 3

AX   P:AJ **90**,115   S:none   Mode 5

BF   P:ApSpSc **78**,141   S:B.A.N **11**,499 (DAO)   Mode 5

BH   P:A&ApSup **42**,195   S:PASP **77**,367   Mode 2

DL   P:A&Ap **61**,107   S:PASP **86**,267   Mode 5

DM   P:AJ **88**,535   S:ApJ **166**,361   Mode 2

Vulpecula

Z   P:A&Ap **61**,469   S:ApJ **126**,53   Mode 5

RS   P:A&Ap **61**,469   S:ApJ **84**,85 and PASP **94**,945   Mode 2

BE   P:A&ApSup **40**,57   S:none   Mode 5

ER   P:A&Ap **238**,145   S:same   Mode 2

HD Catalog

27130   P:AJ **93**,1471   S:same   Mode 2

77581   P:ApJ **226**,264   S:ApJ **235**,570 and A&ApSup **30**,195   Mode -1

149779   P:A&Ap **167**,53   S:none   Mode 2

153919   P:ApJ **226**,264   S:MNRAS **163**,13P   Mode -1

199497   P:ApSpSc **115**,309   S:none   Mode 3

Miscellaneous

SMC X-1 P:ApJ **226**,264   S:ApJ **217**,186   Mode 4

Glossary

The terms defined below include many which may be encountered in further reading about binary stars, as well as those used in the **Pictorial Atlas**.

* * * * * * * * *

Absolute dimensions (of stars): Dimensions (*e.g.* radius, diameter) in a standard length unit such as kilometers or astronomical units.

Absolute orbit: The orbit of a star as described with respect to an inertial (*i.e.* non-accelerated) coordinate system. In a binary or multiple star system this refers to a star's motion with respect to the center of mass of the whole system. The more massive star has the smaller absolute orbit, and the absolute orbit sizes are inversely proportional to the masses. See **relative orbit**.

Albedo (bolometric): The ratio of bolometric radiant energy leaving a surface (whether emitted, scattered, or reflected) to that incident on the surface. The bolometric albedo is 1.00 (all incident energy is returned) if there is local conservation of energy. Stars whose outer envelopes transfer energy mainly by convection may conserve energy only globally and not locally.

Algol-type binary: A binary star system which has experienced large scale mass transfer so as to change the higher mass star into the lower mass star, and is now in a condition of slow, intermittent mass transfer. The transfer is by lobe overflow from a sub-giant star to an un-evolved or nearly un-evolved main sequence star. Large numbers of Algols are known because the condition is fairly long lasting, and because those with orbits at all close to edge-on are easily detected by their deep eclipses.

Apastron: The point on a non-circular orbit around a star which is most distant from the star and where the orbital motion is slowest. In an elliptical orbit, apastron is at one end of the major axis, with **periastron** at the other end.

Balance point: As used here, a point of zero effective gravity on a binary star line of centers. The **inner Lagrangian point**, which pertains only to the case of synchronous rotation in circular orbits, is a special case of the concept of a balance point. A **limiting lobe** contains a **balance point**.

Binary star system: Two stars which are bound together by their gravitation so as to continue to orbit one another.

Black body: An object with an idealized surface which absorbs all incident radiation, and thus is a perfect absorber. Such a surface also will be a perfect emitter of radiation, in the sense that it emits the maximum possible radiant energy per unit area at a given temperature. The emitted intensity is the same in all outward directions (no limb darkening) and the spectral distribution is given by Planck's radiation law (see **Planckian radiation**). The radiation of real stars is only very roughly represented by Planckian radiation, with the correspondence being best for the hottest stars.

Bolometric: This term refers to radiant energy summed over all wavelengths (or frequencies). A star with a particularly large percentage of its energy outside the visible spectrum (say in the ultraviolet or infrared) might be more luminous **bolometrically** than a typical solar-type star, yet could be less luminous in visible light.

Center of mass: For an object or collection of associated objects, the point which follows a straight line path when there are no external forces acting. The object must have rotation for the center of mass to be identified in this way. In a binary star system, the center of mass is always on the line of centers and closer to the star of larger mass, with the ratio of distances from the stars inversely proportional to the ratio of masses. Most methods for determining binary star mass ratios involve one means or another for locating the center of mass.

Close binary star system: A binary system in which the stars are close enough together so that they exchange matter by lobe-overflow at some time in the life of the system. Thus a binary which has not yet gone through mass exchange, but will do so in the future, is considered a **close binary system**. A **close binary** is to be contrasted with a **wide binary**. In looser usage, the term refers to relative closeness (separation of the binary components relative to their size).

Color index: As a rough description, a quantity which expresses color quantitatively. More generally, **color index** measures the local slope of a spectral energy distribution (and can be applied to regions outside the visible band, where **color** is not meaningful). A **color index** is a difference of two stellar magnitudes, measured at two regions of the spectrum. An example is the **B-V color index**, where **B** stands for blue (about 0.44 micron wavelength) and **V** stands for visual (about 0.55 micron). When used for stars, color indices are indicators of surface temperature (more negative **B-V** means hotter), and of reddening by interstellar dust (more positive **B-V** means more reddening). Several effects can cause the **color index** of a binary star system to vary with orbital phase.

Common envelope (in a binary star system): A region shared between the two stars, extending from the surface down to where the stars are physically distinct. If the rotations of the stars are synchronized with the orbital motion, common envelope configurations can be long lasting, as in the **W UMa** binaries. If rotation is non-synchronous, complicated dynamical events are expected to result in large scale loss of matter from the system and rapid shrinkage of the orbit, leading to the production of very small and even extremely small binaries.

Conjunction (of binary star members): For a circular orbit, the geometric condition of the smallest separation in the plane of the sky as one star passes by the other. For an eccentric orbit, this definition is not exactly correct, although nearly so (see a textbook for precise definition).

Contact binary: An alternative term for an **overcontact binary**.

Convective envelope: A zone in the outer part of star in which thermal (heat) energy is transported outward mainly by moving cells (blobs) of gas. This is convective energy transfer. Convective regions also can exist in cores and intermediate zones of stars, according to circumstances. Alternative possibilities are energy transfer by radiation and by conduction.

Detached binary system: A binary system in which both stars are smaller than their limiting lobes.

Doppler effect: A change in the observed frequency of radiation (or, in fact, any periodic phenomenon) due to approach or recession (motion along the line of sight) between source and observer. The shift is to higher frequency for approach and to lower frequency for recession. For speeds which are fairly small compared to that of light, the percent change of frequency is the same as the percent speed, relative to the speed of light. These frequency shifts are extremely important for binary stars because they provide a means to measure the orbital motion quantitatively, and directly in kilometers per second regardless of the distance to the system. For many other kinds of measurements it is necessary to know the distance in order to put the results into directly useful absolute units.

Double contact binary: One in which both stars accurately fill their limiting lobes. This means that the surface of each star coincides with the surface of its limiting lobe. The stars do not touch one another. This can happen when one star (originally not in contact with its limiting lobe) is set into rapid outer rotation by accreted gas from the other star, since rotation reduces the size of a limiting lobe. The matter-giving star would already have been filling its limiting lobe before this occurred.

Eccentricity: The ratio of the distance between the foci of an ellipse to the length of the major axis.

Eccentric orbit: One in which the orbiting object has a variable distance from the center of mass or the other object. In a Newtonian two-body system, the orbits will be ellipses which follow Kepler's laws (see an astronomy textbook). Light curves of eclipsing binaries with eccentric orbits will have their primary and secondary eclipses unequally spaced and of unequal durations, except for certain special orientations. Eclipse depths and shapes also are affected by orbital eccentricity. Radial velocity curves for eccentric orbits have characteristic shapes, from which the orientation of the major axis and eccentricity can be found.

Electromagnetic (EM) radiation: The most familiar type of radiation, including radio, infrared, visible, ultraviolet, X, and gamma radiation. In some contexts the term "light" may be used for **EM** radiation, for brevity. In other contexts, light may mean only visible **EM** radiation, and in still others, it may also include infrared and ultraviolet radiation. All of these "kinds" of radiation are really only one kind, and the names are based on practical considerations, such as how they are normally detected and how they are usually produced. This is made clear by the fact that, in principle, radio radiation can be Doppler-shifted to become gamma radiation, and vice-versa. The particle of **EM** radiation is the photon. Among radiations which are **not EM** radiation are cosmic rays (atomic nuclei and electrons) and neutrinos.

Ellipsoid: A three-dimensional geometrical figure in which all cross-sections are ellipses, and which is characterized by the lengths of its three axes of symmetry. If all three axes have different lengths, the figure is called a tri-axial ellipsoid. Such ellipsoids are used in some models of binary star systems to represent the forms of tidally and rotationally deformed stars. This is a good approximation for small distortions, but not for large ones (see **physical model**).

Ellipsoidal variation: Periodic brightness variation of a tidally distorted star as it orbits, which is partly due to the changing geometrical cross-section facing the observer, partly due to the surface brightness effect associated with gravity darkening, and partly due to limb darkening. Each of these effects contributes in the same sense to ellipsoidal variation, and all with roughly the same dependence on phase. Mainly, the star is dimmest when seen end-on (twice per orbit, at the conjunctions), and brightest when seen "broadside" (also twice per orbit). Thus the light curve maxima are curved in a convex-upward sense. For some binaries with large tidal distortions, the strength of gravity darkening and of limb darkening can be estimated (with some difficulty) from the effect.

Equipotential surface (or just equipotential): A surface on which the potential energy per gram of matter is the same everywhere. In binary star models the potential energy is understood to be the sum of the gravitational and centrifugal potential energies. The concept is most directly applicable to synchronously rotating, circular orbit binaries, although it can be stretched to apply to other cases. There will be an infinity of nested equipotentials in a binary system, one of which should coincide with the surface of one star, while another coincides with the surface of the other star. The surface of an overcontact binary follows one common equipotential. The surface of a limiting lobe coincides with a particular equipotential - one on which a special point exists, at which the effective gravity is zero (see **balance point**). In general, gravity varies over an equipotential, being relatively strong where the adjacent equipotentials are closely spaced and weak where they are far apart.

Evolution (stellar): The aging of stars. As stars age, they develop non-uniform chemical compositions in nested shells, with successively heavier elements on the inside. The aging of massive stars proceeds much faster than that of low-mass stars at all stages of the evolutionary process, and this is a key point for understanding binary star evolution. The theory of binary star evolution begins from that of single stars and introduces ideas on the effects of mass transfer, tidal interactions, and other binary effects. See **evolved star**.

Evolved star: A star in which thermonuclear burning has changed the original chemical composition by converting hydrogen to helium, and later to still heavier elements. Except in very low mass stars, which can keep well-mixed by convection, this results in non-uniform compositions because the nuclear transformation begins at the center and works outward. In normal usage the term **evolved star** is used interchangeably with **chemically non-uniform** star.

Gravity darkening (or brightening): A phenomenon which was predicted by stellar structure theory, and whose effects are observed routinely in very close binaries. The phenomenon is that local bolometric emission of radiation on a star is greater where local gravity is greater. For radiative envelopes this is a direct proportionality, while for convective envelopes the effect is not as strong, but still large enough to have a major effect on light curves of very close binaries. Some persons prefer to call it darkening, while others call it brightening, but it is the same effect.

Inclination (orbital): In the convention used for binary stars, the angle between the plane of the sky and the orbit plane. Thus an edge-on orbit has an inclination (usual symbol is i) of 90°.

Infrared: The part of the electromagnetic spectrum which lies between the longest wavelengths visible to the human eye (*i.e.* "beyond the red") and the shortest radio waves (microwaves). Low temperature stars have a relatively large percentage of their radiation in the infrared, compared to hotter stars, so that - potentially - a cool star companion to a hot star may be most easily observable in the infrared. However, there are practical difficulties with radiation detectors and transmission of the Earth's atmosphere, and the promise of the infrared is only now beginning to be properly realized.

Inner Lagrangian (L₁) point: The point of zero effective gravity on the line of star centers in the **Roche model** (synchronous rotation, circular orbit). The **inner Lagrangian point** is closer to the star of lower mass. It does not coincide with the center of mass unless the masses are equal.

Kepler's third law: In its modern version, "the sum of the masses in a two-body orbiting system is proportional to the cube of the (length of the) semi-major axis and inversely proportional to the square of the orbital period." Kepler found the semi-major axis vs. period part of the relation, while the mass involvement was found later, as a consequence of Newton's laws of motion and gravitation. This law makes possible the determination of absolute masses for binary stars, planets, and even galaxies.

Light curve: A record of brightness vs. time. For close binary systems, time is usually expressed as "phase", for which the unit of time is the orbit period. Light curves for periodic or nearly periodic variations are usually "folded", which means that successive cycles are plotted atop one another. Folding helps one visualize an idealized full cycle of variation when only fragments of individual cycles are available. Brightness may be expressed in stellar magnitudes (a logarithmic measure of brightness) or directly in the amount of light received. In normal practice, the "magnitudes" are actually differences of magnitudes between the star of interest and a (presumed constant) comparison star. If the plot is in terms of light, the light may include an arbitrary scale factor (constant multiplier). The light curves of the **Pictorial Atlas** are in light, not magnitudes, and their scaling is explained in the introductory reading.

Limb darkening: A physical phenomenon in which the intensity of radiation is progressively dimmer toward the limb (edge of visible disk) of a star. It follows from the local situation on the star's surface in which there is a decrease of intensity with angular distance from the vertical. This is expected for a semi-transparent radiating medium whose temperature increases inward, and thus should occur for all normal stellar atmospheres, although it will be a stronger effect for some stars than others. Limb darkening tends to be relatively strong for cool stars, and on a given star it tends to increase toward shorter wavelengths. It affects binary star light curves by increasing the effective ellipsoidal variation, and by changing the shapes of eclipses, especially in the case of annular eclipses which become rounded on the bottom, rather than flat.

Limiting lobe: A closed surface in a binary system which marks the largest one of the stars can be and have the matter on all parts of its surface bound to the star. For a star whose surface coincides with a limiting lobe, the effective gravity is zero at one special point, and matter at that location will be lost. Should the star expand slightly, a hole of finite size will open and a large flow through the hole will necessarily occur, preventing the star from growing larger than the lobe. An exception is possible for the case in which **both** stars reach their lobes, so that neither can accept matter from the other. An **overcontact** binary, with both stars larger than their lobes, is then the expected result.

Luminosity (of a star): The radiant energy emitted per unit time from the entire surface. For a **bolometric luminosity**, the total radiation of all kinds is included, and the unit might be ergs per second or the bolometric luminosity of the Sun. A **monochromatic luminosity** includes only the radiation in a band centered around a specific wavelength, and the unit might be ergs per second per unit wavelength interval, or might be the luminosity of the Sun in that band. Since in practice the wavelength bands for monochromatic luminosities often are not particularly narrow (and certainly not arbitrarily narrow), a better name would have been **polychromatic luminosity**, but that name is almost never used. If there is no adjective and the kind of luminosity is not made clear by the context, then **luminosity** is understood to mean **bolometric luminosity**.

Luminosity class: Classification of a star according to luminosity, based on spectral line strength criteria (see **spectral type**). A star is first assigned a **spectral type**, which represents its surface temperature, and then a **luminosity class**. Criteria for luminosity classification change along the spectral sequence. The actual physical condition measured by luminosity classification is surface gravity. However, at a given temperature, a star of low gravity will normally be more luminous than one of high gravity, so a surface gravity estimate is equivalent to a luminosity estimate. There are six luminosity classes - Ia, Ib, II, III, IV, and V - with Ia being the most luminous supergiants and V being main sequence stars.

Magnitude (stellar): A quantity used in astronomy to represent the apparent brightnesses of stars and other sky objects within some specified region of the spectrum. A difference of magnitudes corresponds to a ratio of received light. Each one magnitude difference represents a brightness ratio of the fifth root of 100 (the irrational number 2.512 . . .), so that a difference of five magnitudes represents exactly a brightness ratio of 100. The magnitude of an object includes also an additive constant which is based on certain adopted magnitudes for standard stars. Eclipsing binary light curves usually are published in the form of **magnitude differences** vs. time, where the differences are between the eclipsing star and a (presumed constant) comparison star. A magnitude "system" can be defined for any region of the spectrum, and many systems are actively used. Each is characterized approximately by an effective wavelength and a bandwidth, except for **bolometric** magnitudes, which include the entire spectrum.

Main sequence: A band which runs from the upper left (high surface temperature, high luminosity) to the lower right (low temperature, low luminosity) of the Hertzsprung-Russell diagram. See an astronomy textbook for an explanation of the Hertzsprung-Russell diagram. In terms of stellar structure, the **main sequence** is the locus of positions of young stars of various masses (and some definite original chemical composition) which still have uniform or only modestly non-uniform compositions. Stars on the main sequence have not yet converted their central hydrogen entirely to helium via thermonuclear burning, and still have roughly their original size and luminosity. Stars spend most (say 90%) of their nuclear burning lifetimes on the main sequence and, because of these long lifetimes, main sequence stars are very common members of binary systems. The luminosity class of a main sequence star is V.

Mass transfer (in binary stars): Exchange of matter between stars. The main mechanisms are **lobe overflow** (see **limiting lobe**) and **stellar winds**. Mass transfer by lobe overflow is understood to be responsible for **Algol** type binaries, in which the matter flow has made the originally more massive star now the less massive one.

Model: An idealized representation of a physical system or object. The models now used for close binary systems include rather large numbers of interrelated features, and automatic computation is required to produce model light curves and other predictions of observable properties.

Morphology of close binary stars: Classification according to overall forms of binaries, in terms of whether stars are smaller, the same size, or larger than limiting lobes. See entries for **detached**, **semi-detached**, **overcontact**, and **double contact** binaries.

Overcontact binary: A binary star system in which both stars are larger than (overfill) their limiting lobes. Overcontact binaries normally are in synchronous rotation and have circular orbits. Their limiting lobes are accordingly call Roche lobes, a term which is appropriate only under those circumstances. The outer part of an overcontact binary - particularly the part which lies above the Roche lobes - is assumed to form an envelope which is shared in common by the two stars (common envelope), and to have one value of potential energy everywhere on its surface. The best known overcontact binaries are the W UMa stars. Overcontact binaries are often simply called **contact** binaries. However, **overcontact** is more accurately descriptive because historical tradition has been to use the term **contact** to mean contact of a star with a limiting lobe, rather than with its companion star.

Parameter: A characteristic quantity (such as mass ratio, orbital inclination, etc.) of a model which has a definite, but unknown, value. The value can often be estimated by analysis of observations or by adoption from theory.

Periastron: The point on a non-circular orbit around a star which is closest to the star, and where the orbital motion is fastest. On an elliptical orbit, **periastron** is at one end of the major axis, with **apastron** at the other end.

Periastron (longitude of): The binary orbit intersects the plane of the sky in a line called the line of nodes. The longitude of periastron (usual symbol = ω) is the angle between that line and the major axis of the orbital ellipse. (See a textbook for a more complete definition). Thus the longitude of periastron describes how the elliptical orbit is oriented within its own plane, with respect to the direction of the observer. In favorable situations, this parameter can be estimated very accurately from binary star light curves, and often also from radial velocity curves.

Period (orbital): The time interval to complete one orbit. It ranges from much less than one day to roughly two weeks for the binaries in the **Pictorial Atlas**. The range for all known eclipsing binaries is from less than 1.5 hours to 27.1 years.

Phase (orbital): Time expressed in the unit of an orbital period. Usually the whole number of cycles is disregarded and only the digits after the decimal place are retained (see entry for **light curve**).

Photometric mass ratio: An estimate of the ratio of masses of the two stars in a binary system which is based on the relation between mass ratio and relative Roche lobe size. If a star fills its Roche lobe accurately, as does one star in a semi-detached binary, the star size tells the lobe size which specifies the mass ratio. Another case is that of an overcontact binary for which the sizes of **both** stars depend on the mass ratio and on the amount of overcontact, but mainly on the mass ratio. In the **Notes** section of the **Pictorial Atlas**, this quantity is represented by q_{ptm}.

Photometric solution: Estimation of values of model parameters by analysis of a light curve. Light curve model parameters are far more numerous than those of radial velocity solutions because eclipses and other photometric effects allow us to probe the stars as resolved objects, rather than only as point sources of light. However, only a subset of the parameters can be reliably estimated for any one binary. The process consists of varying the parameters, usually in some systematic way, until there is a good match between an observed light curve and a computed light curve, based on the model.

Physical model (of a binary star system): This term is used to distinguish models which utilize an equipotential representation of the star surfaces and surface gravities from models which utilize ellipsoids (geometrical models).

Planckian radiation: The radiation of an idealized **black body** as it depends on temperature and wavelength (or frequency). The intensity of radiation is given by a relatively simple expression derived by M. Planck (see an astronomy textbook). Although stars are only roughly approximated by Planckian radiators, their radiative properties are often calibrated in terms of Planckian radiators. For example the **effective temperature** of a star is the temperature that a Planckian radiator would have if it produced the same radiation flux (per unit surface area) as the star.

Plane of sky: A plane at right angles to the line of sight from the observer to a sky object and passing through the object.

Primary eclipse: The deeper of two eclipses (*i.e.* the one of greater loss in system brightness) in an eclipsing binary. In binaries which have circular orbits, the deeper eclipse is nearly always that of the hotter star. It is not particularly unusual, however, to have the star of **lower** luminosity covered at the primary eclipse, even when the orbit is circular. That can happen if the lower luminosity star is the hotter and smaller star, which can be the case, for example, in **RS CVn** type binaries.

Primary star: The main star in a binary system according to some adopted convention. A spectroscopist usually will choose the more massive star, unless the mass ratio is unknown, in which case it will be the star with the more prominent spectral absorption lines. A photometric observer usually will choose the star with deeper eclipses.

Radial velocity curve: A record of radial (line of sight) velocity vs. time. For some binaries, the spectral lines of both stars can be measured from Doppler shifts, and we have a double-lined binary, with information on the motions of both stars. More often, one of the stars is too faint to impress measurable lines on the spectrum and we have a single-lined binary. Radial velocity and light curves provide complementary information about a binary system, and both are needed to form a reasonably complete description.

Radiative envelope: A zone in the outer part of a star in which thermal (heat) energy is transported mainly by electromagnetic radiation. This is radiative energy transfer. Radiative zones also can exist in other parts of stars, according to circumstances. Alternative possibilities are transfer by convection and by conduction.

Reflection effect: Increased radiative brightness on the side of a star which faces toward a nearby companion star. This is mainly due to heating by the radiant energy of the other star, with subsequent re-emission of the energy at locally raised temperatures, so the effect is not literally one of reflection. However, in very hot stars a considerable fraction of the incident radiation is simply scattered by free electrons, so in that case the term "reflection effect" is somewhat more appropriate. One effect of reflection on binary star light curves is to raise the brightness level around the secondary eclipse relative to that near the primary (deeper) eclipse. Another is to produce a concave-upward curvature between eclipses (curvature opposite to that from ellipsoidal variation).

Relative dimensions (of stars): Dimensions (*i.e.* radius, diameter) of binary member stars, with the semi-major axis of the relative orbital ellipse as the unit. Thus a star in a circular orbit whose relative radius is 0.22 has a radius 22 percent as large as its distance from the other star.

Relative orbit: The orbit of one star in a binary system with respect to the other star. The **relative orbits** of the two stars have, of course, the same size, which is the size of the two **absolute orbits** added together.

Roche lobe: A **limiting lobe** for the case of synchronous rotation and a circular orbit.

Roche model: An idealization of a binary star system which has somewhat different meanings to different persons. It is characterized by the specification of a binary system in terms of **equipotential surfaces** (level surfaces) under the assumptions of synchronous rotation, circular orbits, and gravitation arising entirely from point masses at the centers of the stars. The forms of star surfaces, magnitude and direction of local gravity, and other physical properties can then be computed from the properties of the equipotential surfaces. Other models are in use in which some or all of these assumptions are replaced by more general ones. The term **Roche model** is then not appropriate, but **extended Roche model** may be used.

Rotation: Angular motion with respect to an axis. When rotation proceeds uniformly in time (essentially the case for stars, at least over short intervals) it is characterized by a period of rotation. Rotation of a star affects its figure (producing polar flattening) and surface effective gravity (progressively reducing the effective gravity away from the poles, with minimum values at the equator).

RS Canum Venaticorum (RS CVn) type binary: Member of a class of detached close binaries recognized as a class only recently (1960's) but now known to be abundant, interesting, and important to the understanding of binary star structure and evolution. **RS CVn**'s contain two stars of nearly equal mass and show effects of magnetic starspots, chromospheric activity, and active circum-binary regions. In a normal **RS CVn** binary, the star of slightly larger mass is the star of larger size and lower surface temperature. Many **RS CVn**'s have been detected as radio sources and most or all show brightness variation due to spots as the spotted star rotates. Such effects make it fairly easy to identify **RS CVn**'s even when the orbital inclinations are too low to produce eclipses, and some of the very interesting ones are not eclipsing binaries. Many other characteristics of **RS CVn** binaries are very noteworthy, but too numerous to mention here.

Semi-detached binary: A binary system in which one star is detached and the other in contact with (said to be "filling") its **limiting lobe**. A more directly descriptive term would be **semi-contact**, but the potentially confusing term **semi-detached** is the one in use. The best known type of semi-detached binary is the **Algol** class.

Semi-major axis: Half the length of the major (longer) axis of an ellipse. If the ellipse is a circle (*i.e.* has zero eccentricity) the semi-major axis is equivalent to the radius.

Spectral type: A classification of a star into one of seven main types and their decimal subtypes according to surface temperature. From hot to cool, the types are O, B, A, F, G, K, and M. The decimal subtypes result in designations such as A0 (the hottest A star), G5 (middle of the G range), and K9 (the coolest K star). There are a few additional types such as R and N (for stars with unusually high abundances of carbon), and W (for Wolf-Rayet stars, which are very unusual in several ways). Classification is based on relative strengths of certain standard pairs of absorption lines, with the pairs to be used being different at different sub-ranges of the spectral sequence.

Spectroscopic mass ratio: An estimate of the ratio of masses of the stars in a double-lined binary system, based on the relative amplitudes of the two radial velocity curves. The spectroscopic mass ratio estimate is essentially equal to the inverse of the ratio of velocity amplitudes (*i.e.* $m_2/m_1 = K_1/K_2$, where K is the standard symbol for velocity curve amplitude), although for stars with large tidal distortions or reflection, it may be advisable to account for those effects. In the **Notes** section of the **Pictorial Atlas** this quantity is represented by q_{rv}.

Spectroscopic solution: Estimation of values of model parameters by analysis of radial velocity observations. The parameters are the orbital period, eccentricity, longitude of periastron, time of periastron passage, and the product of the length of the semi-major axis and the sine of the orbital inclination ($a \sin i$). See **photometric solution**.

Spot (accretion): A hot spot on a star or on a disk around a star which is the target of a fairly narrow stream of incoming gas. Such narrow streams can occur as a result of lobe overflow (see **limiting lobe**) in semi-detached binaries. However, in many examples, the stream first misses the target star, is accreted later after it has become diffuse, and does not produce a well-defined localized hot spot.

Spot (magnetic): A low temperature region of the same nature as a sunspot, resulting from penetration of a star's surface (photosphere) by a bundle of magnetic field lines. According to present theory and observations, these dark spots occur only on stars with outer convective envelopes. Among binaries which show such spots are the **RS CVn** stars, on which they are often extremely strong and produce major effects on light curves, and **W UMa** stars.

Sub-giant star: A star which has evolved beyond the main sequence stage, but has not yet become a giant star. It is therefore intermediate in size between a main sequence star and a giant star. Its thermonuclear burning is hydrogen burning in a shell around a non-burning, nearly pure helium core. The luminosity class is IV. The secondary star of a typical **Algol** binary is a **sub-giant**.

Synchronous rotation (of an orbiting object): Rotational behavior in which the rate of rotation is synchronized with orbital angular motion. The simplest case is that of equal rotational and orbital angular rates in a circular orbit. The rotational and orbital periods will, of course, then be the same. Most very close binaries rotate in this way because tides have circularized their orbits and locked together their rotational and orbital motion. In eccentric orbits, tidal theory predicts that a fluid body such as a star should synchronize its rotation at a rate which is constant over the orbit and very nearly equal to the orbital angular rate at periastron. Other kinds of synchronous rotation can occur for solid bodies, but not for normal stars.

Systemic velocity: The velocity of the center of mass of a physical system. For a binary star system, the term traditionally refers to the radial (line of sight) velocity of the center of mass, which can be estimated from radial velocities of either star separately, or from the velocities of both stars. The usual symbol for systemic velocity is V_γ

Ultraviolet (UV): The part of the radiation spectrum between the shortest wavelengths visible to the human eye and the X-ray region. High temperature stars have a relatively large percentage of their radiation in the ultraviolet, compared to cooler stars, and are therefore very prominent in **UV** observations. The **far UV** is the spectral region which is blocked by the Earth's atmosphere and can only be observed from space.

Wide binary: A binary system in which the stars are so widely separated that neither surface will reach its limiting lobe, even in the most expanded evolutionary state, so that mass transfer by lobe overflow cannot occur. Accordingly, the two stars will exert little influence on one another and go through their "lives" essentially as two single stars. See **close binary**.

W Ursae Majoris (W UMa) type binary: An overcontact binary of low surface temperature (not exceeding about 8000 K), short period (less than 3/4 of a day), low mass (total usually a few solar masses or less), and mass ratio far from unity (from about 0.7 to 0.07). These are main sequence stars with common convective outer envelopes. The class is subdivided into the W-type and the A-type W UMa binaries. Present evidence indicates that the W-types are at a very early evolutionary stage (essentially "zero-age"), while the A-types are somewhat more massive binaries which have evolved significantly beyond "zero-age", although still on the main sequence. **W UMa**'s are very abundant binaries which have attracted much observational and theoretical attention, and are very controversial in regard to their origin, structure, and evolution.

Index by Constellation

Coma Berenices
   RW, 9
   RZ, 34
Corona Australis
   TZ, 126
Corona Borealis
   U, 281
   RT, 211
   RW, 81
   α, 334
Corvus
   W, 35
   RV, 82
Cygnus
   Y, 312
   SW, 212
   UZ, 335
   VW, 313
   WW, 282
   BR, 127
   CG, 36
   MR, 157
   MY, 283
   V382, 314
   V388, 128
   V442, 158
   V453, 315
   V463, 159
   V470, 284
   V478, 316
   V548, 160
   V1073, 83
   V1143, 285
   V1425, 161
   V1727, 10
Delphinus
   W, 286
   DM, 37
Draco
   RZ, 38
   UZ, 213
   AI, 129
   AX, 84
   BH, 162
   BS, 214
   BV, 85

   BW, 11
   CM, 86
Equuleus
   S, 215
Eridanus
   RU, 87
   WX, 88
   YY, 39
   AS, 163
   BW, 89
   CW, 216
Gemini
   RW, 287
   RY, 317
   UX, 336
   YY, 90
   AF, 130
   GW, 91
Grus
   W, 217
Hercules
   RX, 164
   SZ, 131
   TT, 132
   TX, 165
   UX, 166
   AD, 318
   AK, 40
   HS, 219
   LT, 133
   V338, 134
   V624, 220
   u, 218
Horologium
   SY, 41
Hydra
   TT, 288
   VZ, 167
   WY, 92
   AI, 319
   HS, 168
   KM, 289
   χ^2, 221

Hydrus
   Y, 222

Indus
   RS, 93
   RY, 94
Lacerta
   RT, 223
   SW, 42
   VY, 135
   AR, 169
   AW, 224
   CM, 170
   CO, 171
   EM, 43
Leo
   UV, 95
   XY, 12
   AM, 44
Leo Minor
   T, 290
Libra
   ES, 96
   δ, 225
Lupus
   FT, 45
Lynx
   RR, 320
   SW, 97
   UU, 46
   UV, 47
Lyra
   TZ, 98
   FL, 172
   V361, 48
Mensa
   TZ, 321
   UX, 226
Monoceros
   RU, 291
   RW, 173
   TU, 322
   VV, 292
   AO, 227
   AR, 337
   AU, 338
   IM, 174
Musca
   TU, 293